Ionizing Radiation - Insights into Chemistry, Biology, and Applications

Edited by Haitham Sghaier

Published in London, United Kingdom

Ionizing Radiation - Insights into Chemistry, Biology, and Applications
http://dx.doi.org/10.5772/intechopen.1004664
Edited by Haitham Sghaier

Contributors
Alexandra Bran, Ambaye Fisehaw Tesfaw, Anca Filimon, Anca Roseanu, Cherie Sisay Mekonnen, Cristian Munteanu, Daniel Avram, Eckehard Mueller, Felix Sima, Gabriela Chiritoiu, Gabriela Negroiu, Ioana Porosnicu, Ion Tiseanu, Jeremiah Odhek Masime, Joseph Owuor Lalah, Kais Manai, Livia Sima, Madalina Icriverzi, Monica Tudor, Omrane Kadri, Paula Florian, Peter Munyao Ndangili, Sandel Simion, Stefana Orobeti, Wubshet Getachew Mengesha, Yohichi Kohzuki

First published in London, United Kingdom, 2025 by IntechOpen
IntechOpen is the global imprint of INTECHOPEN LIMITED, registered in England and Wales, registration number: 11086078, 167-169 Great Portland Street, London, W1W 5PF, United Kingdom

For EU product safety concerns: IN TECH d.o.o., Prolaz Marije Krucifikse Kozulić 3, 51000 Rijeka, Croatia, info@intechopen.com or visit our website at intechopen.com.

British Library Cataloguing-in-Publication Data
A catalogue record for this book is available from the British Library

Ionizing Radiation - Insights into Chemistry, Biology, and Applications
Edited by Haitham Sghaier
p. cm.
Print ISBN 978-1-83634-059-1
Online ISBN 978-1-83634-058-4
eBook (PDF) ISBN 978-1-83634-060-7

If disposing of this product, please recycle the paper responsibly.

Meet the editor

Prof. Dr. Ir. Haitham Sghaier is an Engineer in Biotechnology and a Professor (Full) of Biological Engineering (Cellular and Molecular Bioengineering) / Radiation & Computational Microbiology at the National Center for Nuclear Sciences and Technology (CNSTN) in Tunisia. In 2016, he won the TWAS Young Arab Scientist (YAS) Prize. Prof. Sghaier is the Founding President of the Tunisian Association for Computational Sciences (TACoS) and the Founding Project Chief of the Joint Research Platform for the Fight against the Emerging Coronavirus (COVID-19) at CNSTN/BiotechPole Sidi Thabet (from Nov 09, 2020, to Mar 21, 2022) in Tunisia. Currently, he is coordinating two projects, funded by the International Atomic Energy Agency (IAEA), that focus on the development of irradiated vaccines.

Contents

Preface IX

Chapter 1 1
Advanced Computations and Simulations of Gamma-Ray Buildup Factors in Human Tissues and Tissue-Equivalent Materials
by Omrane Kadri and Kais Manai

Chapter 2 19
Radiation Caused by Elementary Particle
by Eckehard Mueller

Chapter 3 29
New Peak on Thermoluminescence Glow Curve of X-Ray-Irradiated Single Crystal after Heat Treatment
by Yohichi Kohzuki

Chapter 4 43
Advancements of Nanomaterials for Diagnostic Imaging, Radiotherapy, and Nuclear Medicine
by Wubshet Getachew Mengesha, Cherie Sisay Mekonnen and Ambaye Fisehaw Tesfaw

Chapter 5 57
X-Ray Techniques in Analytical Chemistry
by Jeremiah Odhek Masime, Peter Munyao Ndangili and Joseph Owuor Lalah

Chapter 6 79
Ultra-High Dose Rate Ionizing Radiation for FLASH Radiotherapy Applications
by Livia Sima, Monica Tudor, Stefana Orobeti, Anca Filimon, Gabriela Negroiu, Cristian Munteanu, Gabriela Chiritoiu, Anca Roseanu, Paula Florian, Madalina Icriverzi, Ioana Porosnicu, Daniel Avram, Alexandra Bran, Sandel Simion, Ion Tiseanu and Felix Sima

Preface

This Edited Volume is a collection of reviewed and relevant research chapters concerning the developments in ionizing radiation. The book includes scholarly contributions by various authors and was edited by a group of experts knowledgeable about ionizing radiation. Each contribution comes as a separate chapter, complete in itself but directly related to the book's topics and objectives.

The target audience comprises scholars and specialists in the field.

IntechOpen

Chapter 1

Advanced Computations and Simulations of Gamma-Ray Buildup Factors in Human Tissues and Tissue-Equivalent Materials

Omrane Kadri and Kais Manai

Abstract

This chapter provides a comprehensive examination of advanced computations and simulations of gamma-ray buildup factors in human tissues and tissue-equivalent materials, pivotal for applications in medical physics, radiation therapy, and diagnostic imaging. Gamma-ray interactions with matter, particularly human tissues, are complex phenomena that significantly impact the accuracy of dosimetry and imaging techniques. The chapter delves into photon energy absorption and exposure buildup factors specific to human tissues, presenting key methodologies and findings from recent experimental studies. Utilizing sophisticated simulation tools such as Geant4, we model gamma-ray transmission through various tissue-equivalent materials, providing detailed insights into their interaction characteristics. These simulations are validated against experimental data, ensuring robustness and reliability. We explore the search for optimal tissue-equivalent materials, focusing on their potential to enhance the precision of radiation dose calculations. By integrating computational and simulation approaches, this chapter elucidates the role of buildup factors in improving the accuracy of medical imaging modalities and in optimizing radiation therapy protocols. The implications of our findings extend to the development of new materials and technologies in radiation shielding and detection. Through this in-depth analysis, we aim to advance the understanding of gamma-ray interactions, contributing to safer and more effective medical treatments and diagnostic procedures.

Keywords: gamma-ray buildup factors, human tissues, tissue-equivalent materials, Geant4 simulation, medical physics, radiation dosimetry, medical imaging

1. Introduction

Gamma rays play a critical role in medical imaging and radiation therapy worldwide, yet the potential dangers associated with their improper use underscore the necessity for a thorough understanding of radiation interaction with human tissues. The precise interaction of ionizing radiation with biological tissues varies depending on several factors, including the type and energy of the radiation as well as the target tissue. One key parameter in understanding these interactions is the buildup factor (BUF), which corrects for the limitations of the Beer-Lambert law by accounting for the scattering and attenuation of photons within a material, as follows (see **Figure 1**) [1]:

$$I = BI_0 e^{-\mu x} \tag{1}$$

where I, I_0, B, μ, and x are transmitted and incident beam intensities, buildup factor, linear attenuation coefficient, and attenuator thickness, respectively. On the other hand, tissue-equivalent materials (TEMs) are essential in replicating the behavior of human tissues under radiation, providing a valuable tool in medical applications such as radiation therapy and diagnostic radiology. These materials must closely mimic the absorption and scattering characteristics of real tissues, and the BUF is involved in evaluating their effectiveness. Although various methods, including Monte Carlo simulations, invariant embedding, and geometric progression (GP) fitting, have been used to study gamma-ray buildup factors, the most precise dataset currently available is from the ANSI/ANS-6.4.3-1991 standard, which offers the GP fitting parameters for a range of elements and compounds [2]. However, despite the availability of this dataset, there remains a need for comprehensive data on BUFs for human tissues at deep penetration levels. Moreover, the Geometry ANd Tracking (Geant4) Monte Carlo toolkit [3] offers a promising approach to simulate these factors accurately. In this study, we aim to expand the existing knowledge by providing a database of BUFs for various human tissues, simulated using Geant4, and computed by GP fitting method. Our research seeks to address several critical areas: first, the verification of the calculation methods against established standards; second, an exploration of BUF-dependent parameters such as energy and penetration depth; and third, the evaluation of TEMs as potential substitutes for human tissues in radiation applications. Through a series of computations, we calculated BUFs for some human tissues and examined their applicability to point kernel-based programs used in high-definition reference human models such as the HDRK-Man [4].

Within this context, it was verified that the geometric parameter "K" behaves linearly with buildup factors for energies up to 15 MeV and short depths of 8 mfp leading to reducing the GP fitting parameters to three only. Within a standard deviation of no more than 1%, BUF were reproduced by the new fitting method [5]. On the other hand, it was found that SB3, glycerol trioleate, MS15, AP6, and stracey latex are

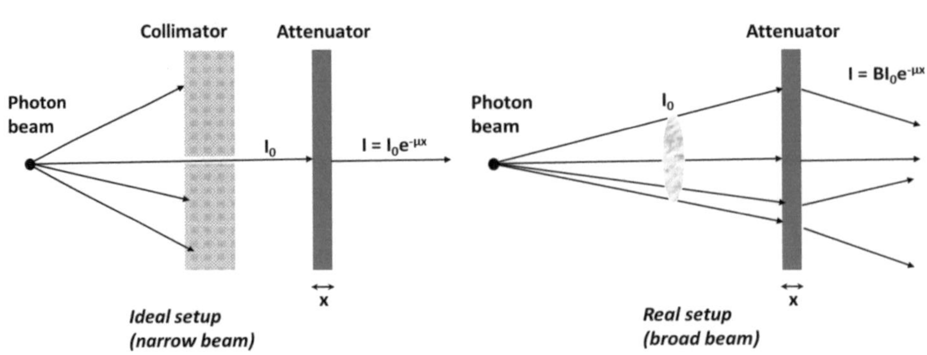

Figure 1.
Schematic view describing the situation of using the buildup factor for realistic setup compared to the idealized one of a monochromatic narrow photon beam traveling through a thin layer of attenuator.

acceptable tissue material variants for cortical bone, fat, muscle, adipose, and lung tissues, respectively [6]. We can consider the actual findings as an extendion work of our previous ones. In this chapter, we will present key findings on photon energy absorption and exposure in human tissues. Then, we will explore the criteria for selecting tissue-equivalent materials based on the computation of exposure and energy absorption buildup factors. Additionally, we will discuss the results of Geant4 simulations related to gamma-ray transmission buildup factors in water medium. Finally, we will outline the implications, future directions, and emerging trends in buildup factor research within the fields of medical diagnostics and treatment.

By introducing a new tuning parameter for the TEM selection process—based on the BUF—we aim to refine the criteria for selecting equivalent materials in gamma-ray applications. This study not only advances the field of radiation physics by offering a more accurate simulation framework but also provides valuable insights for ongoing research in therapeutic and diagnostic radiology.

2. Photon energy absorption and exposure in human tissues

This section provides the computational methodology adopted to compute buildup factors and the application for HDRK-man tissues and organs. For medical imaging purposes, we limited our study to the photon energy range covering the 15–150 keV interval, with nine points, for penetrations up to 40 mean free path (mfp), with 17 values. Meanwhile, our developed C^{++} code goes until 15 MeV as photon energy.

2.1 Computation methodology

To determine the buildup factors, the GP fitting parameters were derived through interpolation based on the equivalent atomic number (Z_{eq}), which is an energy-dependent parameter that characterizes the properties of composite materials in terms of equivalent elements. The computational process is detailed step by step as follows:

- Calculation of the equivalent atomic number.
- Determination of geometric progression (GP) fitting parameters.
- Computation of buildup factor values (energy absorption buildup factor [EABUF] and exposure buildup factor [EBUF]).

The process begins by extracting the Compton partial mass-attenuation coefficient $(\mu/\rho)_{Compton}$ and the total mass attenuation coefficients $(\mu/\rho)_{Total}$ for elements with atomic numbers ranging from Z = 4 to 92, as well as for selected dosimetric materials in the energy range of 0.015–15 MeV, using the WinXCom software [7] (originally developed as XCOM). To interpolate Z_{eq}, for which the ratio $R = (\mu/\rho)_{Compton}/(\mu/\rho)_{Total}$ of a given human organ or tissue falls between the corresponding ratios of two successive elements (with $R1$ and $R2$), the following semi-logarithmic formula is applied [8]:

$$Y = \alpha \log(X) + \beta \tag{2}$$

with X and Y are the ratio R and the Z_{eq}.α and β are the segment equation parameters determined with two points (X1,Y1) and (X2,Y2).

Then, using the GP fitting parameters for pure elements from the ANSI/ANS-6.4.3-1991 standard reference and the previously calculated Z_{eq}, the corresponding GP fitting parameters (a, b, c, d, and Xd) for the buildup factors are interpolated using the same semi-logarithmic interpolation formula [9–11], with X and Y are Z_{eq} and any of the five parameters. Finally, the GP parameters employed for calculating the energy absorption buildup factor (EABUF) and the exposure buildup factor (EBUF) differ for each combination of energy and material. Therefore, all relevant parameters from the ANS-6.4.3-1991 standard were first tabulated. The buildup factors for a given incident photon energy E and penetration depth x in mean free path units were then calculated using the following approach [5, 6, 12, 13]:

$$B(E,x) = 1 + (b-1) \times \begin{cases} x & if \quad K = 1 \\ \dfrac{K^x - 1}{K - 1} & if \quad K \neq 1 \end{cases} \tag{3}$$

where, for x ≤ 40 mfp:

$$K(E,x) = cx^a + d\frac{\tanh(x/X_d - 2) - \tanh(-2)}{1 - \tanh(-2)} \tag{4}$$

and for x ≥ 40 mfp:

$$K(E,x) = \begin{cases} 1 + (K_{35} - 1)\Phi^{\xi(x)} & if \qquad 0 \leq \Phi \leq 1 \\ K_{35}\left(\frac{K_{40}}{K_{35}}\right)^{0.8\xi(x)} & otherwise \end{cases} \tag{5}$$

where, $\Phi = \frac{K_{40}-1}{K_{35}-1}$, $\xi(x) = \frac{(x/35)^{0.1}-1}{(40/35)^{0.1}-1}$, $K_{35} = K(E,35)$ and $K_{40} = K(E,40)$.

For deep penetration scenarios exceeding 40 mfp, the extrapolation of the dose multiplication factor K(E, x) as proposed by Harima et al. [14] and supported by Sakamoto and Trubey [15] was applied.

On the other hand, as we aim to compute buildup factors for HDRK-man tissues and organs, **Table 1** illustrates their elemental composition and density (in g/cm^3). A C^{++}-based program was developed for that purpose and running on Lunix Operating System.

2.2 Key findings

Figure 2 shows the computed buildup factors ((left) EABUF and (right) EBUF) for brain tissue at different photon energies, function of the mean free path. We can see the closeness of both types of BUF to each other, which is due to the low energy interval. Moreover, the difference between them will be clear for more than 1 MeV, where the pair-production process becomes dominant despite the current case where the dominance of the Compton effect exists and begins slowly decreasing against the photoelectric effect, confirming literature and theory for all three processes.

Material	Density (g/cm^3)	Elemental composition[a]
Adipose	0.92	H(11.40), C(59.80), O(27.80)
Adrenal	1.02	H(10.50), C(25.60), N(2.70), O(60.20)
Blood	1.06	H(10.20), C(11.00), N(3.30), O(74.50)
Bone	1.34	H(7.34), C(25.48), N(3.06), O(47.89)
Brain	1.03	H(10.7), C(14.5), N(2.2), O(71.2)
Breast	0.92	H(11.40), C(59.80), O(27.80)
Colon	1.04	H(10.60), C(11.50), N(2.20), O(75.10)
Colon[b]	1.03	H(10.60), C(11.50), N(2.20), O(75.10)
ET	1.03	H(10.50), C(25.60), N(2.70), O(60.20)
Gall	1.03	H(10.50), C(25.60), N(2.70), O(60.20)
Gall[b]	1.03	H(10.50), C(25.60), N(2.70), O(60.20)
Heart	1.03	H(10.40), C(13.90), N(2.90), O(71.80)
Heart[b]	1.06	H(10.20), C(11.00), N(3.30), O(74.50)
Intestine	1.04	H(10.60), C(11.50), N(2.20), O(75.10)
Intestine[b]	1.03	H(10.60), C(11.50), N(2.20), O(75.10)
Kidneys	1.05	H(10.30), C(13.20), N(3.00), O(72.40)
Liver	1.06	H(10.20), C(13.90), N(3.00), O(71.60)
Lung	0.30	H(10.13), C(10.24), N(2.87), O(75.75)
Muscle	1.04	H(10.20), C(14.30), N(3.40), O(71.00)
Esophagus	1.04	H(10.50), C(25.60), N(2.70), O(60.20)
Esophagus[b]	1.04	H(10.50), C(25.60), N(2.70), O(60.20)
Oral mucosa	1.03	H(8.43), C(57.40), N(1.61), O(24.55)
Pancreas	1.06	H(10.60), C(16.90), N(2.20), O(69.40)
RBM	1.03	H(10.50), C(41.40), N(3.40), O(43.90)
Salivary glands	1.03	H(10.50), C(25.60), N(2.70), O(60.20)
Skin	1.09	H(10.00), C(20.40), N(4.20), O(64.50)
Spleen	1.06	H(10.30), C(11.30), N(3.20), O(74.10)
Thymus	1.03	H(10.50), C(25.60), N(2.70), O(60.20)
Thyroid	1.05	H(10.40), C(11.90), N(2.40), O(74.50)

[a]*Elements with percentage less than 1% are omitted (can be found elsewhere) [4].*
[b]*Tissue content.*

Table 1.
Elemental composition and density (in g/cm^3) of different HDRK-man studied tissues and organs.

Also, the other radiological properties of the studied case of "brain tissue," such as the equivalent atomic (Z_{eq}), mass-attenuation coefficients (μ/ρ and μ_{en}/ρ), and the five GP-based fitting parameters ($a, b, c, d,$ and Xd) of buildup factors, were tabulated in **Table 2**. All these parameters will be used for searching a material mimicking the brain tissue.

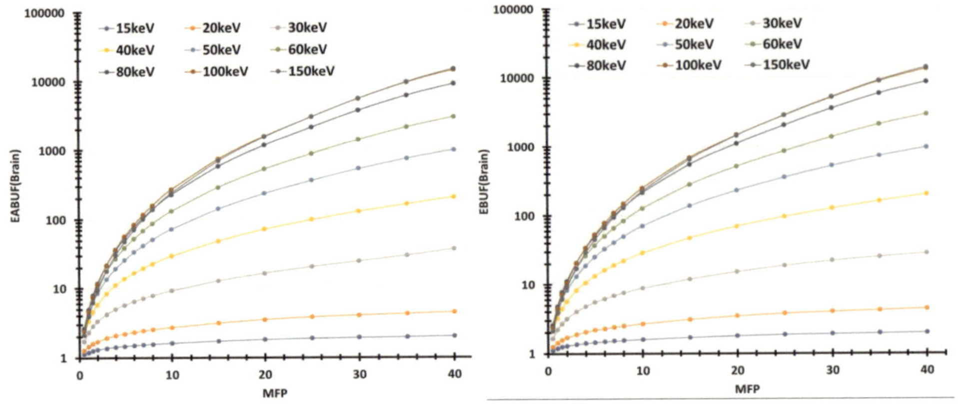

Figure 2.
Computed buildup factors ((left) EABUF and (right) EBUF) for brain tissue at different photon energy, function of mean free path.

3. Search for tissue-equivalent materials

This section summarizes some criteria describing the physical properties of 30 known and usual materials for selecting the closest one able to behave for this photon energy range, similarly to brain tissue similarly to brain tissue (**Table 3**).

3.1 Criteria for selecting tissue-equivalent materials

The mixture rule adopted for the calculation of the mass-attenuation coefficient for a given material is given by:

$$\frac{\mu}{\rho} = \sum_i w_i \left(\frac{\mu}{\rho}\right)_i \tag{6}$$

where the fractional weight of the ith element, of the material, figured with w_i.

The effective atomic number, Z_{eff}, is a crucial parameter in assessing tissue equivalence, radiation absorption, scattering, and shielding effectiveness for gamma and neutron radiation in composite materials [16]. Among the various methods available for determining the effective atomic number, the direct computational method was employed for the selected tissue substitutes using a specific formula [6, 17]:

$$Z_{eff} = \frac{\sum_i f_i A_i \left(\frac{\mu}{\rho}\right)_i}{\sum_i f_i \frac{A_i}{Z_i} \left(\frac{\mu}{\rho}\right)_i} \tag{7}$$

where, f_i and A_i denote the molar fraction and the atomic mass of the ith element.

Additionally, the effective electron density, N_{eff}, which represents the number of electrons per unit mass, is directly proportional to Z_{eff} as follows:

$$N_{eff} = N_A \frac{Z_{eff}}{\sum_i n_i A_i} \tag{8}$$

E	Z_{eq}	μ/ρ	μ_{en}/ρ	EABUF					EBUF				
(keV)		(cm^2/g)	(cm^2/g)	a	b	c	d	Xd	a	b	c	d	Xd
15	7.458	1.710	1.412	0.180	1.197	0.457	−0.090	13.896	0.180	1.196	0.455	−0.091	14.374
20	7.542	0.828	0.570	0.151	1.448	0.536	−0.076	14.660	0.148	1.438	0.541	−0.074	14.805
30	7.696	0.381	0.163	0.097	2.329	0.711	−0.044	12.848	0.076	2.218	0.754	−0.044	16.179
40	7.805	0.270	0.073	0.000	3.360	1.042	−0.008	13.335	−0.000	3.286	1.043	−0.007	13.474
50	7.867	0.227	0.044	−0.065	4.259	1.353	0.024	13.479	−0.063	4.186	1.347	0.022	13.579
60	7.909	0.206	0.033	−0.107	4.774	1.606	0.046	13.679	−0.105	4.687	1.594	0.044	13.744
80	7.960	0.183	0.026	−0.151	4.942	1.924	0.068	13.700	−0.148	4.807	1.902	0.065	13.714
100	7.991	0.170	0.026	−0.170	4.628	2.083	0.077	13.716	−0.164	4.498	2.045	0.072	13.854
150	8.037	0.150	0.028	−0.169	3.923	2.112	0.069	14.387	−0.162	3.867	2.062	0.063	14.483

Table 2.
Computed equivalent atomic (Z_{eq}), *mass-attenuation coefficients* (μ/ρ *and* μ_{en}/ρ), *and the five GP-based fitting parameters (a, b, c, d, and Xd) of buildup factors (EABUF and EBUF) for brain tissue.*

Material	Density (g/cm^3)	Elemental composition
Brain (ICRU)	1.040	H(12.001), C(64.005), N(0.800), O(22.902), Na(0.050), Mg(0.002)
BRN1	0.920	H(8.359), C(69.133), N(2.360), O(16.938), F(3.070), Cl(0.140)
BRN2	0.930	H(11.180), C(53.310), O(35.510)
BRN3	0.920	H(14.370), C(85.630)
BRN4	2.700	Al(100.000)
BRN5	1.350	H(4.840), C(38.436), Cl(56.724)
BRN6	1.840	H(3.100), C(31.260), N(0.990), O(37.570), Cl(0.050), Ca(27.030)
BRN7	2.100	C(24.020), F(75.980)
BRN8	1.720	H(4.686), O(56.630), P(10.867), K(27.817)
BRN9	0.910	H(14.369), C(85.607), O(0.004), Sb(0.020)
BRN10	0.950	H(11.701), Li(3.48)), C(72.748), O(12.071)
BRN11	0.920	H(11.840), C(77.320), O(10.840)
BRN12	0.320	H(5.741), C(73.947), N(2.010), O(18.142), Sb(0.160)
BRN13	0.3 00	H(6.000), C(51.440), N(4.290), O(30.720), Al(7.550)
BRN14	0.260	H(10.100), B(8)), C(79.200), O(0.120), S(1.910), Zn(0.670)
BRN15	1.000	H(8.830), C(64.450), N(4.050), O(20.350), Cl(2.240), Sb(0.080)
BRN16	1.000	H(8.871), C(66.817), N(3.100), O(21.132), Sb(0.080)
BRN17	1.250	H(5.740), C(77.460), O(16.800)
BRN18	1.070	H(10.200), C(12.010), N(3.540), O(74.250)
BRN19	1.200	H(5.549), C(75.573), O(18.878)
BRN20	1.050	H(11.430), C(65.580), O(9.220), Mg(13.480), Ca(0.290)
BRN21	0.990	H(13.401), C(77.799), O(3.500), Mg(3.860), Ti(1.440)
BRN22	1.000	H(9.750), C(63.160), N(0.940), O(16.020), Al(9.600), Cl(0.530)
BRN23	1.000	H(8.119), C(58.344), N(1.780), O(18.638), Mg(13.029), Cl(0.090)
BRN24	1.400	H(9.799), C(63.683), N(12.379), O(14.139)
BRN25	0.930	H(14.860), C(85.140)
BRN26	1.170	H(8.051), C(59.986), O(31.963)
BRN27	1.050	H(7.74), C(92.26)
BRN28	1.120	H(10.2), C(76.8), N(3.6), O(5.9), F(1.7), Ca(1.8)
BRN29	1.010	H(9.654), C(87.489), N(0.06), O(0.473), S(1.539) Ti(0.332), Zn (0.453)
BRN30	1.000	H(11.19), O(88.81)

Table 3.
Percentage elemental composition and density of 30 studied tissue-equivalent candidates.

where, the Avogadro number and the number of atoms are given by N_A and n_i, respectively.

Another key parameter for comparing the radiation characteristics of tissue and tissue-equivalent materials is the CT index or Hounsfield Unit (H.U.), defined by [6]:

$$H.U. = \frac{\langle \mu \rangle - \langle \mu_{water} \rangle}{\langle \mu_{water} \rangle - \langle \mu_{air} \rangle} \tag{9}$$

where $\langle \mu_{air} \rangle$, $\langle \mu_{water} \rangle$, and $\langle \mu \rangle$ are the average linear attenuation coefficients for air, water, and material in concern, respectively. The H.U. values are typically calculated for a standard photon beam spectrum at 120 kVp. In this study, the standard 120 kVp spectra were generated using the SpeckCalc program [18].

3.2 Key findings

For the studied photon energy range, **Figure 3** shows the calculated equivalent atomic number for brain tissue (straight line) and the corresponding possibly equivalent material.

Also, the comparison to brain tissue in terms of buildup factor (**Figure 4**), mass-attenuation coefficient (**Figure 5**), and mass-energy attenuation coefficient

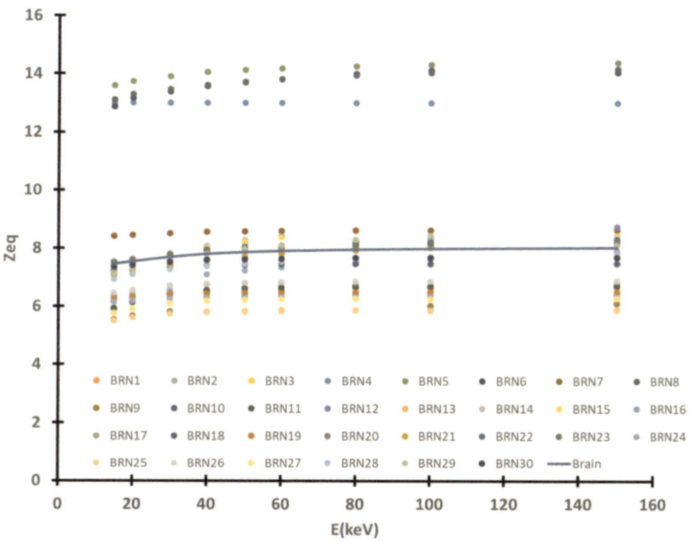

Figure 3.
Calculated equivalent atomic number (Z_{eq}) function of photon energy, up to 150 keV, for brain tissue (straight line) and 30 equivalent material candidates (dotted curves).

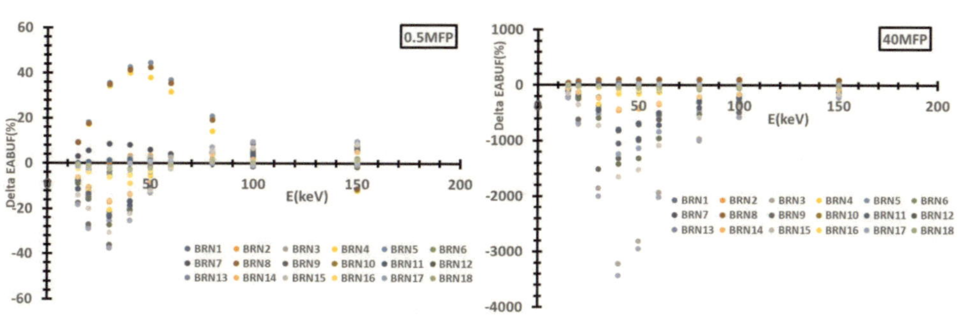

Figure 4.
Relative deviation (Delta in %) of 30 equivalent material candidates to brain tissue, in terms of EABUF for extrema values of the studied mfp interval (0.5–40 mfp).

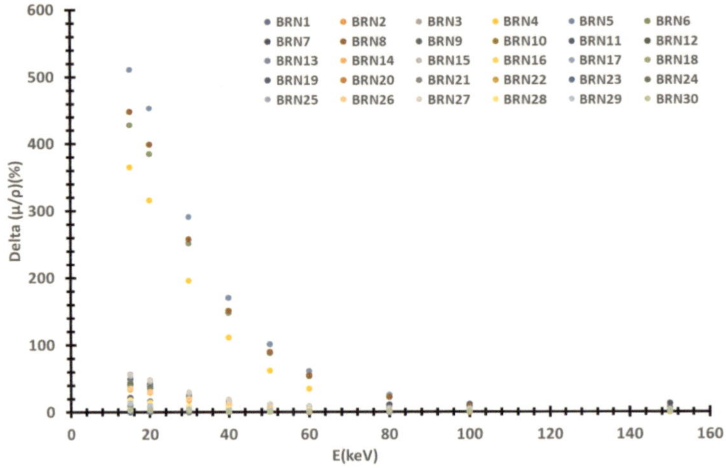

Figure 5.
Relative deviation (Delta in %) of 30 equivalent material candidates to brain tissue in terms of mass-attenuation coefficient (μ/ρ) *for photon energy between 15 and 150 keV.*

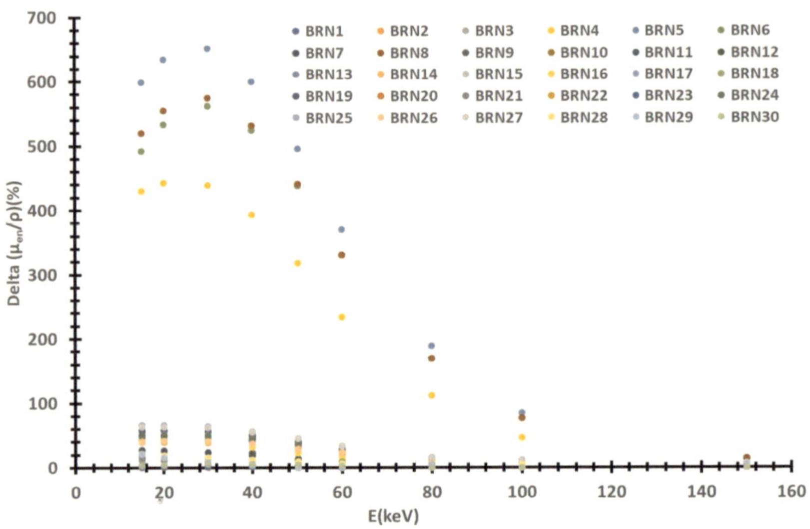

Figure 6.
Relative deviation (Delta in %) of 30 equivalent material candidates to brain tissue in terms of mass-energy attenuation coefficient (μ_{en}/ρ) *for photon energy between 15 and 150 keV.*

(**Figure 6**) were shown for all the 30 candidates. Those relative deviation (in %) were summarized in **Table 4**. Moreover, the corresponding difference of H.U. values, against H.U. (brain), were tabulated.

Firstly, we can confirm the theoretical hypothesis of the inverse proportionality relationship between buildup factors and Z_{eq}. From all of the above results, we can conclude that the material *BRN*20 is the most material mimicking brain tissue. Therefore, we can conclude that “M3” is an almost ideal tissue-equivalent material (TEM) for brain tissue, making it particularly suitable for medical dosimetry, radiotherapy, and imaging applications, especially in head diagnosis.

4. Geant4 simulations of gamma-ray transmission

In this study, Geant4 version 10.4 was employed to simulate the transport of gamma rays emitted from a point source through water. The geometry modeled consisted of a monoenergetic point source positioned at the center of concentric spheres. We simulated the isotropic transport of 0.1 MeV photons generated from the center of an infinite sphere with an outer radius of 105.45 cm, corresponding to 18 mean free paths (mfp). Using energy flux-to-exposure conversion factors from the literature [19], the simulated exit flux (measured at each step of one mfp) was converted into the exposure buildup factor, utilizing the mass-energy absorption coefficient of air obtained from the NIST website [20]. The exposure buildup factor (EBF) for water at 0.1 MeV photon energy was calculated up to a penetration depth of 10 mfp. The simulation involved 10^8 gamma rays, with bremsstrahlung processes deactivated and all other electron and photon processes activated. We used the "G4EmStandardPhysics_option3" physics library with a 1 keV cutoff for electrons and photons. The overall statistical uncertainty for all simulations did not exceed 1%.

Table 5 demonstrates good agreement between our current computations and results published in the literature [21] for the exposure buildup factor in a water medium, considering a point isotropic source of 0.1 MeV photons with penetrations up to 10 mean free paths (mfp). In our simulations, we employed the "discrimination by energy" method with a 10% energy window as described elsewhere [22]. The minor discrepancies observed, particularly at greater penetration depths, could be addressed by either utilizing a supercomputer or by applying particle splitting as a variance reduction technique [21, 23]. These discrepancies are likely due to the reduced probability of photon detection as the distance from the source increases.

5. Applications in medical imaging and radiation therapy

Even this present study can be considered important and valuable for physician researchers; however, the study of more physical properties of TEM candidates, including elasticity, manufacturing, and others should be carried out.

5.1 Implications of buildup factor studies in medical diagnostics and treatment

Buildup factor studies play a crucial role in enhancing the accuracy and effectiveness of medical diagnostics and radiation therapy. In medical imaging, understanding the buildup factors allows for better dose estimation, which is vital for minimizing patient exposure while ensuring high-quality imaging. For radiation therapy, accurate buildup factor data contributes to precise dose delivery to the target tissues, reducing the risk to surrounding healthy tissues. Moreover, buildup factors could be integrated into imaging modalities such as CT, MRI, and PET scans and their impact on treatment planning in radiotherapy, ensuring optimized therapeutic outcomes are straightforward.

5.2 Future directions and emerging trends

As medical imaging and radiation therapy continue to evolve, the study of buildup factors is expected to advance in parallel. Emerging trends may include the

Material	EABUF	EBUF	Z_{eq}	Z_{eff}	μ/ρ	μ_{en}/ρ	N_{eff}	Average	H.U. (120kVp)
BRN1	126	173	16	7	15	29	13	54	15
BRN2	89	117	14	13	12	25	5	39	55
BRN3	389	585	26	28	18	39	8	156	13
BRN4	66	70	67	215	123	268	98	129	−530
BRN5	69	72	80	123	181	404	28	137	−2397
BRN6	68	71	75	137	154	349	60	131	−1577
BRN7	20	22	10	98	11	16	61	34	−3558
BRN8	68	71	75	125	159	357	35	127	−1785
BRN9	227	306	21	28	18	37	8	92	−43
BRN10	134	177	17	20	17	35	8	58	−19
BRN11	132	175	17	20	17	35	8	58	−41
BRN12	30	32	8	16	13	20	27	21	−44
BRN13	4	4	1	19	4	3	21	8	−55
BRN14	8	9	3	6	2	6	21	8	−85
BRN15	12	14	4	3	5	8	14	9	−99
BRN16	55	61	9	6	13	24	13	26	−159
BRN17	176	250	19	13	18	33	24	76	−10
BRN18	24	31	5	3	4	9	3	11	−39
BRN19	156	219	18	14	18	32	24	69	−50
BRN20	**2**	**3**	**1**	**6**	**0**	**1**	**4**	**2**	**−25**
BRN21	9	10	2	13	3	5	5	7	−92
BRN22	6	7	2	2	1	2	8	4	−84
BRN23	5	6	2	9	2	1	12	5	−94
BRN24	196	282	20	11	16	31	10	81	157
BRN25	405	613	26	30	18	39	8	163	−258
BRN26	86	113	13	7	14	26	13	39	−35
BRN27	230	327	21	10	21	40	19	96	−196
BRN28	22	26	5	5	6	10	13	12	−7
BRN29	6	7	3	4	4	6	20	7	−92
BRN30	12	17	3	5	1	3	6	7	−78

Table 4.
Relative deviation, for photon energy 15–150 keV and depth up to 40 mfp, of studied material candidates and brain tissue properties, in terms of BUFs, equivalent and effective atomic number (Z_{eq} and Z_{eff}). The mean values and H.U. for all of them were also given. Notice that BRN20 material was the closest one for brain tissue mimicking.

development of more sophisticated computational models and simulation techniques, incorporating artificial intelligence (AI) to predict and analyze buildup factors more accurately. Moreover, such a technique could be useful for the TEM selection

MFP	Geant4	Hirayama	Delta (%)
1	4.30	4.55	5.49
2	11.17	11.80	5.31
3	22.92	23.80	3.68
4	39.89	41.30	3.42
5	62.90	65.20	3.54
6	92.84	96.70	3.99
7	130.15	137.00	5.00
8	177.79	187.00	4.93
10	301.90	321.00	5.95

Table 5.
Comparison of Geant4 simulated EABUFs to those of Hirayama for water medium attenuating 0.1 MeV photon beam.

procedure among a huge number of tissue-mimicking material candidates. Furthermore, the increasing use of personalized medicine, tailoring treatment to individual patient characteristics, may also drive the need for more precise and individualized buildup factor data. The potential of these innovations to further improve accurate and efficient diagnosis, as well as the integration of novel materials and technologies in medical applications, is obvious.

6. Conclusions

This chapter has provided an in-depth analysis of advanced computational techniques and simulations used to assess gamma-ray buildup factors in human tissues and tissue-equivalent materials. By highlighting the complicated nature of gamma-ray interactions with matter, we underscore the critical importance of accurate dosimetry and imaging in the fields of medical physics, radiation therapy, and diagnostic imaging. Our exploration of photon energy absorption and exposure buildup factors specific to human tissues reveals significant insights into the methodologies and experimental findings initiating this area of research. The utilization of sophisticated simulation tools such as Geant4 has enabled us to model gamma-ray transmission through water medium, yielding detailed insights into their interaction characteristics. Validation of these simulations against experimental data ensures their robustness, reinforcing their applicability in clinical settings. Moreover, the quest for optimal tissue-equivalent materials is essential for enhancing the precision of radiation dose calculations, ultimately improving patient outcomes. By integrating computational and simulation approaches, we have explained the crucial role of buildup factors in refining medical imaging modalities and optimizing radiation therapy protocols. The implications of our findings extend the immediate applications to the development of innovative materials and technologies in radiation shielding and detection. As we advance our understanding of gamma-ray interactions, we contribute to the ongoing efforts to enhance the safety and efficacy of medical treatments and diagnostic procedures. Further advancements, fostering a more profound understanding of radiation interactions and their implications for human health, could be carried out for future research in this field.

Abbreviations

GP	geometric progression
BUF	buildup factor
EABF	energy absorption buildup factor
EBF	exposure buildup factor
BRN1	AP6
BRN2	ethoxyethanol
BRN3	polyethylene
BRN4	aluminum
BRN5	P.V.C.
BRN6	SB3
BRN7	teflon
BRN8	Witt liquid
BRN9	alderson fat
BRN10	FT1
BRN11	glycerol trioleate
BRN12	alderson lung
BRN13	LN1
BRN14	stacey latex
BRN15	alderson muscle 1
BRN16	alderson muscle 2
BRN17	bakelite
BRN18	goodman liquid
BRN19	lexan
BRN20	M3
BRN21	mix D
BRN22	MS15
BRN23	MS20
BRN24	nylon-6
BRN25	paraffin wax
BRN26	perspex
BRN27	polystyrene
BRN28	shonka plastic
BRN29	temex
BRN30	water

Author details

Omrane Kadri[1*†] and Kais Manai[2†]

1 Research Laboratory on Energy and Matter for Nuclear Science Development (LR16CNSTN02), National Center for Nuclear Science and Technologies, Tunis, Tunisia

2 Physics Department, Faculty of Sciences Tunis - El Manar II, Tunis, Tunisia

*Address all correspondence to: okadri@cern.ch

† These authors contributed equally.

References

[1] Harima Y. An historical review and current status of buildup factor calculations and applications. Radiation Physics and Chemistry. 1993;**41**:631-672. DOI: 10.1016/0969-806X(93)90317-N

[2] American Nuclear Society. Gamma Ray Attenuation Coefficient and Buildup Factors for Engineering Materials. Report ANSI/ANS-6.4.3. USA: American Nuclear Society; 1991

[3] Agostinelli S, Allison J, Amako K, et al. Geant4- a simulation toolkit. Nuclear Instruments and Methods in Physics Research Section A: Accelerators, Spectrometers, Detectors and Associated Equipment. 2003;**506**(3):250-303. DOI: 10.1016/S0168-9002(03)01368-8

[4] Kim CH, Choi SH, Jeong JH, et al. HDRK-man: A whole-body voxel model based on high-resolution color slice images of a Korean adult male cadaver. Physics in Medicine and Biology. 2008; **53**(15):4093-4106. DOI: 10.1088/0031-9155/53/15/006

[5] Alfuraih A, Kadri O. Geant4 simulation of gamma-ray transmission buildup factors for human tissues. Dose Response. 2022;**20**(1): 15593258211068621. DOI: 10.1177/15593258211068621

[6] Kadri O, Alfuraih A. Search for tissue equivalent materials based on exposure and energy absorption buildup factor computations. Applied Sciences. 2022;**12**: 798. DOI: 10.3390/app12020798

[7] Berger MJ, Hubbell JH. XCOM: Photon Cross Sections on a Personal Computer. NBSIR 87–3597. Washington, USA: National Bureau of Standards; 1987. DOI: 10.2172/6016002

[8] Kurudirek MM, Ozdemir Y. Energy absorption and exposure buildup factors for some polymers and tissue substitute materials: Photon energy, penetration 380 depth and chemical composition dependence. Journal of Radiological Protection. 2011;**31**:117-128. DOI: 10.1088/0952-4746/31/1/008

[9] Harima Y. An approximation of gamma-ray buildup factors by modified geometrical progression. Nuclear Science and Engineering. 1983;**83**(2):299-309. DOI: 10.13182/NSE83-A18222

[10] Maron MJ. Numerical Analysis: A Practical Approach. New York: Macmillan; 1987

[11] Manohara SR, Hanagodimath SM, Gerward L. Energy absorption buildup factors of human organs and tissues at energies and penetration depths relevant for radiotherapy and diagnostics. Journal of Applied Clinical Medical Physics. 2011;**12**(4):296-312. DOI: 10.1120/jacmp.v12i4.3557

[12] Olarinoye IO. Photon buildup factors for some tissues and phantom materials for penetration depths up to 100 MFP. Journal of Nuclear Research and Development. 2017;**13**:57-67

[13] Singh VP, Medhat ME, Badiger NM. Assessment of exposure buildup factors of some oxide dispersion strengthened steels applied in modern nuclear engineering and designs. Nuclear Engineering and Design. 2014;**270**: 90-100. DOI: 10.1016/j.nucengdes.2013.12.046

[14] Harima Y, Tanaka S, Sakamoto Y, et al. Development of new gamma- ray buildup factor and application of shielding calculations. Nuclear Science and Engineering. 1991;**94**:24-35

[15] Sakamoto Y, Trubey DK. Radiation safety information computation center

data package DLC-129/ANS643, Geometric Prog. Gamma-Ray Buildup Factor Coefficients. 1991

[16] Singh VP, Badiger NM. Effective atomic numbers of some tissue substitutes by different methods: A comparative study. Journal of Medical Physics. 2014;**39**(1):24-31

[17] Manohara SR, Hanagodimath SM, Thind KS, Gerward L. On the effective atomic number and electron density: A comprehensive set of formulas for all types of materials and energies above 1 keV. Nuclear Instruments and Methods in Physics Research Section B: Beam Interactions with Materials and Atoms. 2008;**266**:3906-3912

[18] Poludniowski G et al. SpekCalc: A program to calculate photon spectra from tungsten anode x-ray tubes. Physics in Medicine and Biology. 2009; **54**:N433

[19] Smith JH. Flux to dose rate conversion factors for gamma ray exposure (No. CONF-46-63; AED-Conf-63-048-87; JPL-TR-32-439). California Inst. of Tech., Pasadena, CA (United States). Jet Propulsion Lab. 1963

[20] National Institute of Standards and Technology (NIST). X-ray mass attenuation coefficients [Internet]. 2024. Available from: https://www.nist.gov/pml/x-ray-mass-attenuation-coefficients [Accessed: August 23, 2024]

[21] Hirayama HJ. Calculation of gamma-ray exposure buildup factors up to 40 mfp using the EGS4 Monte Carlo code with a particle splitting. Journal of Nuclear Science and Technology. 1995; **32**(12):1201-1207. DOI: 10.3327/jnst.32.1201

[22] Jarrah I, Radaideh MI, Kozlowski T, et al. Determination and validation of photon energy absorption buildup factor in human tissues using Monte Carlo simulation. Radiation Physics and Chemistry. 2019;**160**:15-25. DOI: 10.1016/j.radphyschem.2019.03.008

[23] Rafiei MM, Tavakoli-Anbaran H. Calculation of the exposure buildup factors for x-ray photons with continuous energy spectrum using Monte Carlo code. Journal of Radiological Protection. 2018;**38**:207. DOI: 10.1088/1361-6498/aa9cfa

Chapter 2

Radiation Caused by Elementary Particle

Eckehard Mueller

Abstract

Radiation that does not come from the Earth itself accounts for about a third of all natural radiation that humans are exposed to. Its main source is the Sun, which emits a lot of particles and electromagnetic radiation. The effects of this radiation, known as the solar wind, on the Earth's atmosphere and surface are shown here. The second source is the very high energy radiation coming from intergalactic space. The radiation from these two sources is described, as is its interaction with other nuclei. For a better understanding of the mechanism, the characteristics of the particle structure are explained. The distribution of the different elementary particle and radiation reaching the surface of the Earth are given. Examples of the main interactions are explained like the relative high neutron rate on the surface or why high-energy muons reaching us. A not negligible number of particles reaching the surface are produced by interactions in the ionosphere or atmosphere.

Keywords: cosmic radiation, elementary particles, solar wind, X-rays, interactions

1. Introduction

Natural radiation is caused by different sources all over the world [1]. For example, one source is radon coming out of the earth or radioactive carbon captured by trees and afterward in wood. It is used to determine the age of the Egyptian pharos ([2], p. 1288). Another big source is particles coming to Earth from space ([3, 4], pp. 23–32). Most of the time the particle stream or X-rays from space interacts with the nuclei of atoms. This is the main subject of this chapter. Artificial radiation is not considered. Also, radiation from the atomic shell, such as changing the energy level of electrons, is not covered here.

First, the basics of elementary particles are given for a better understanding. Then, the main sources of particles from outer space and the sun are described, followed by the main particle interactions and their results at the Earth's surface where the human beings live.

2. Basics to understand particle reactions

2.1 Different elementary "components"

Elementary means that this particle cannot be divided into sub-particles. There may be a substructure that exists but cannot be separated. The following **Figure 1** shows all the components for elementary particles:

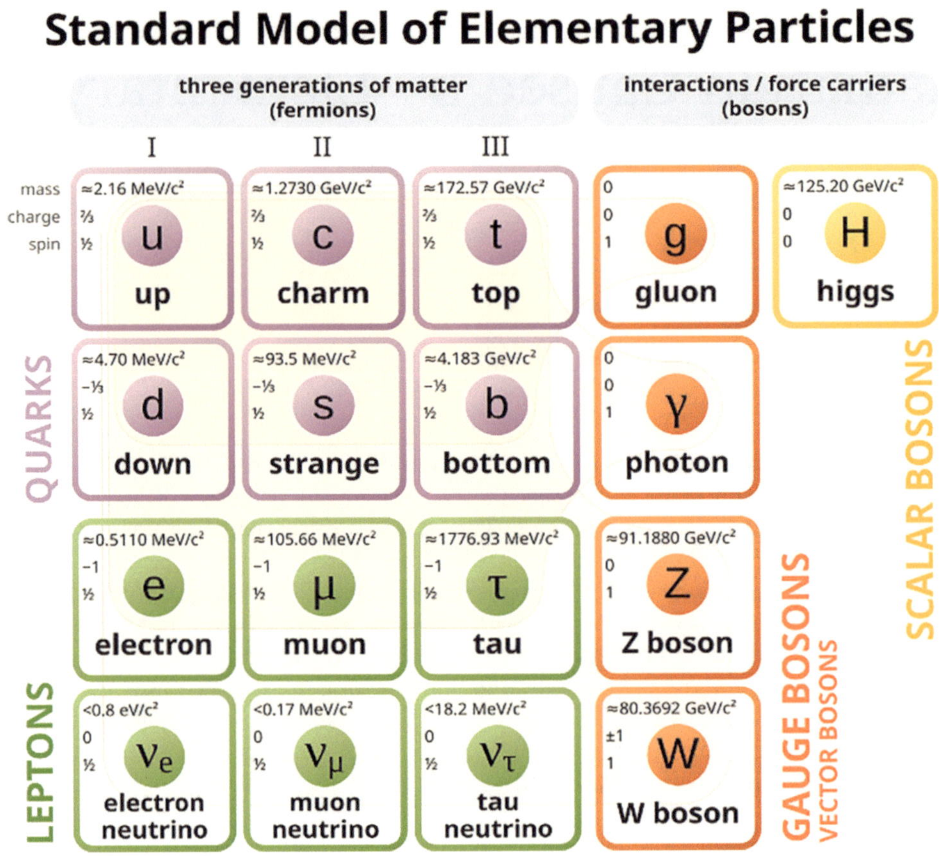

Figure 1.
The different elementary components of matter with their interaction particles (Available from: https://en.wikipedia.org/wiki/File:Standard_Model_of_Elementary_Particles.svg; from cush CC-BY 3.0).

First, we look at the leptons. They are stand-alone particles and never appear as a single particle with anyone else. The electron is the best-known lepton. The muon has the same properties and has about 200 times the mass. The tau plays a minor role and will not be described in detail. If leptons are created every time two particles are created, the associated (anti-)neutrino will pair up with an electron, muon, or tau. It based on the conservation law of the leptonic number ([2], p. 1315), which is one of the four conservation laws in nature.

On the other hand, there are quarks. They never appear alone, which is due to the strong interaction (see Chapter 2.2). The minimum number is two quarks. This particle is called a meson. If there are three quarks, it is a baryon, like the proton (uud) or the neutron (udd) (there are also particles with a higher number of quarks, but these are irrelevant for this topic). It should be noted that the quarks have an elementary charge of +2/3 e or − 1/3 e (elementary charge) e = elementary.

For every lepton and quark, there is an anti-particle with the mirrored properties. The interacting particle is shown on the right (see Chapter 2.2). There is also a conservation law for baryons, which states that the baryon number must be conserved [1].

2.2 Different interactions

There are four different interactions in nature: gravity, electromagnetic interaction, strong interaction, and weak interaction. (The weak and electromagnetic interactions are unified at high energies. At this stage it is called the electroweak interaction).

In this context, only the strong and weak interactions are relevant. Each interaction has a particle. You can call it a force carrier. An interaction field such as the electromagnetic field can be transformed by quantum mechanics into particles, in this case the photon.

The strong interaction has the property, which may be strange to many readers, that the force increases as the distance between two quarks increases. This means that more and more energy has to be put into the process of separating two quarks. When the total amount of energy reaches a certain level, two new quarks are created. This is due to the equivalent between mass m and energy E: $E = m\ c^2$ (c = light speed).

2.3 Definition of an elementary particle

There are different definitions of elementary particles. A restrictive definition is a lepton or a quark. Here is a less restrictive understanding: An elementary particle is a lepton or the combination of at most three quarks. This means that a proton or a neutron are also elementary particles, but not a hydrogen atom, because it is a combination of a lepton (electron) and a baryon, caused by electrostatic attraction.

3. Main natural sources of elementary particles

The most important particle interactions are caused by cosmic. There are two different sources. One is the Sun itself, with the so-called solar wind. An eruption on the Sun results in a burst of particles and electromagnetic radiation reaching the Earth. An indicator of an eruption is the aurora borealis. But also, from interstellar space [5–7] or from galaxies where are other sources emitting X-rays, γ-rays (bursts) [8–10], and particles reaching the Earth, where the source can be identified and others where the scientific community is investigating to identify the origin. The Space Weather Prediction Center of NOAA (National Oceanic and Atmospheric Administration) is an organization that monitors the solar flux. This is a good source of information on solar activity and particle and X-ray showers from the Sun (available from: https://www.swpc.noaa.gov).

Irrespective of the source, cosmic ray showers are now explained. An incident primary particle (usually electron e-, muon μ-, or proton p) has interacted with the nucleus of an atom, or rather with a proton or neutron in the nucleus. The neutrinos are negligible because the cross section is so small that almost no interaction takes place. If you take 7 earths in a line, half of the neutrinos will have interacted. A neutron cannot be an incident particle because it decays in 8 minutes (half-life) into $n -> p + e^- + \overline{\upsilon_e}$. The travel time (speed 300–800 km/s) from the Sun to the Earth is so much longer than the half-life period that no neutron reaches the Earth, but the interaction in the atmosphere produces many neutrons.

All particles interact with the nuclei in an air molecule. The induced shower of particles obtained (photons are also considered as particles) can be divided into three groups [11]:

- leptonic particles and X-rays with low energy (soft section)
- baryons and mesons that mostly decay in further particles (hard section)
- whole nucleons or fragments (nucleonic section)

An example of a shower induced by a proton is shown in **Figure 2** below. For more examples of showers, see [12].

The energy aspect of the incident particles is another important fact. To put it in a nutshell: The higher the energy, the bigger the shower. The energy of the incident particles goes up to 10^{21} GeV [5, 15–16]. The flux is proportional to the energy. (see **Figure 3**).

If the incoming particle is an electron, a weak interaction takes place. A proton is transformed into a neutron. Secondary particles and/or hard X-rays (γ-quants) are produced to get into a lower energy state in the nucleus. The X-ray can interact with a nucleus and produce an electron-anti-electron pair. There could also be an elastic collision or a capture of the electron by the atom. In the end, there is usually an inelastic interaction as described above.

One main particle arriving the earth are protons. The proton interacts with another proton or neutron of the nucleus of an atom. Strong interaction plays the dominant roll. That means interaction between quarks. This causes new particles like the mesons (two-quark particles) π^+ ($u\bar{d}$), π^0 (a quantomechanical superposition $u\bar{u}$ and $d\bar{d}$), or π^- ($d\bar{u}$).) π^+ ($u\bar{d}$), π^0 (a quantomechanical superposition $u\bar{u}$ and $d\bar{d}$), or π^- ($d\bar{u}$).

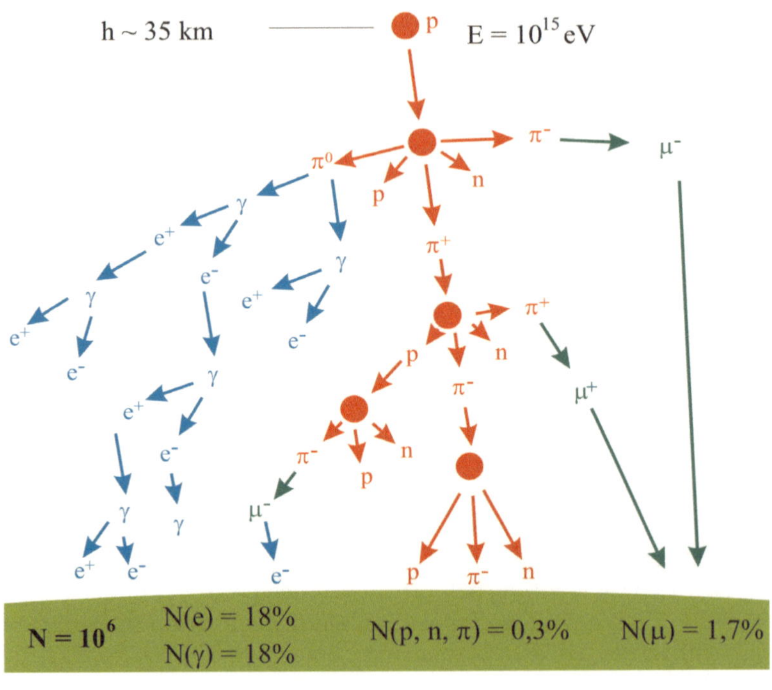

Figure 2.
A typical shower of a proton with the three different sections: blue and green = soft section, red big dots = nucleonic section, red further particles = hard section) (source: https://commons.wikimedia.org/wiki/File:AirShower.svg; from Mpfiz, GNU-License) [12–14].

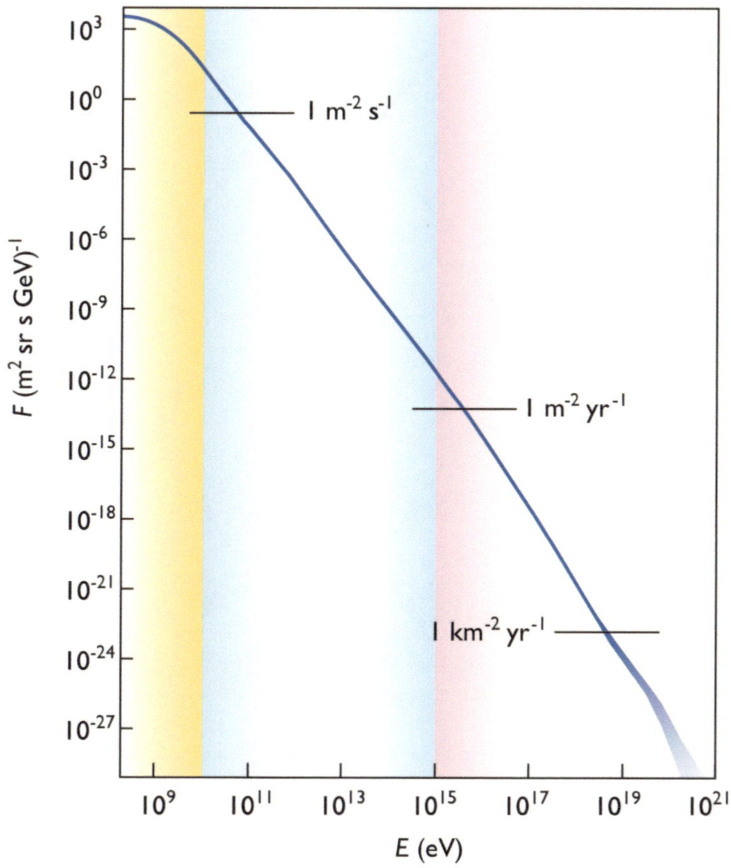

Figure 3.
The flux in dependence of the particle energy for energies greater 10^9 GeV (Available from: https://commons.wikimedia.org/wiki/File:Cosmic_ray_flux_versus_particle_energy.svg; from Sven Lafebre, CC BY-SA 3.0) ([4], p. 577).

4. Reactions and the different results

This chapter describes some of the main reactions in detail. There are more, but that would go beyond the scope of this chapter.

A pion π^+ decays by weak interaction into a muon μ^+ and a neutrino (The μ^+ is the antiparticle of μ^-. It has a positive charge but the same properties. $\bar{d}$ is the antiquark of d). The direction of the arrows is for an antiparticle in the opposite direction. This is not important for us. The u quark and $\bar{d}$ quark annihilate to form a W^+ Boson (force carrier) and this decays into an antimuon and a neutrino (**Figure 4**). The interaction with the π- meson is shown in red (**Figure 5**).

This is a typical reaction that produced μ^- leptons that can reach the surface of the Earth. Also, cosmic muons and electrons [17] coming from other spaces with very high energy will penetrate the atmosphere and be stopped in various regions of the Earth.

There are many different interactions that cascade as shown in **Figure 2**, and particles reach the Earth's surface. Describing them all is beyond the scope of this article. At least four particles reach the surface: protons, neutrons, electrons, and muons.

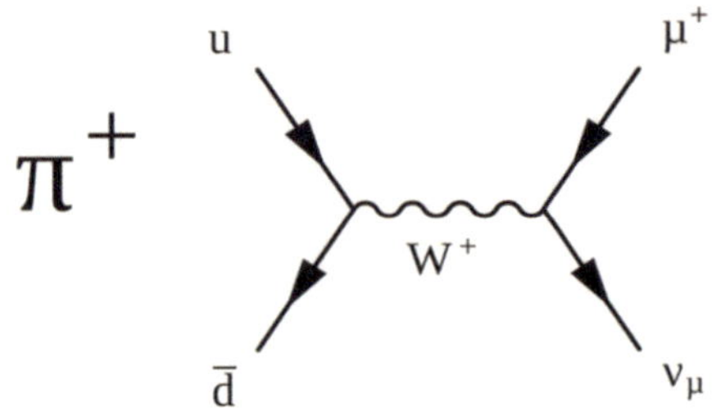

Figure 4.
Feynman diagram of a pion+ decay.

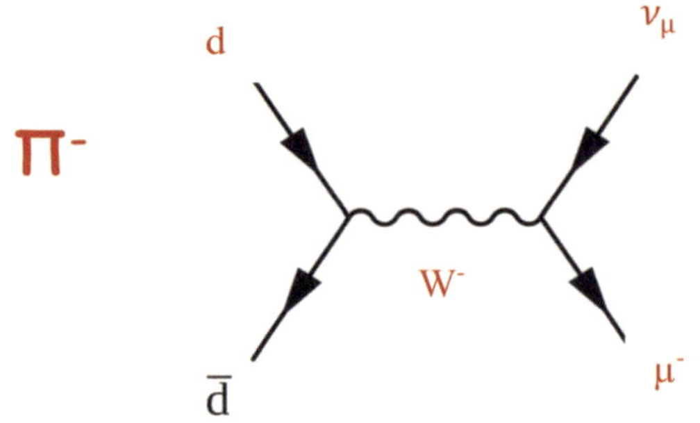

Figure 5.
Feynman diagram of a pion- decay.

5. Conclusions

Different types of particles and electromagnetic radiation coming from outside the Earth affect the amount of radiation on the Earth. There are two main sources, the Sun, and space. The Sun is the dominant source, producing lots of protons and electrons, which are also carried into the atmosphere in the so-called solar wind. Within our atmosphere, many reactions are possible, producing at least four relevant particles: protons, neutrons, muons, and electrons (neutrinos have a very small cross section.) Normally, three particles have a charge of $+/-\ e^{-}$. This is also the cause of the aurora, which is particularly visible around the maximum of the solar cycle, which will be reached in 2025. The interstellar muons coupled with gamma rays have high energy and will hit the Earth. They may scatter in the atmosphere. The further away from the Earth's surface, the higher the intensity of all particles. The atmosphere and the magnetic field act as a shield for charged particles. The total radiation exposure is 0.7 mSv per year [18], which is about 1/3 of the total natural radiation received by a person. Compared to the amount of other radiation today, it is less, but it contributes to evolution [19]. The radiation could also affect electronic parts such as notebooks. [13, 20]. For example, a reboot has to be done. A change of the radiation mix will not be changing the next hundreds of years. The sun works in the 11 years turn from maximum-to-maximum radiation very constant of hundred thousand of years. Also, the intergalactic space in normally quieted or constant, with the exception of Super Novas that affect nearly not out earth because of the great distances.

Author details

Eckehard Mueller[1,2]

1 Bochum University of Applied Sciences, Bochum, Germany

2 Steinbeis-Transfercenter for Spring Technologies, Component Behavior and Process, Iserlohn, Germany

*Address all correspondence to: eckehard.mueller@hs-bochum.de

References

[1] Demtröder W. Experimentalphysik 4. Heidelberg: Springer Spektrum; 2017. p. 39

[2] Tipler PA, Mosca G, Kersten P, editors. Tipler Physik. Heidelberg: Springer Spektrum; 2014. p. 1288

[3] Hanslmeier A. Astronomie. Berlin: Springer Spektrum; 2021. p. 187

[4] de Angelis A, Pimenta M. Introduction to Particle and Astroparticle Physics (Multimessenger Astronomy and its Particle Physics Foundations). Heidelberg: Springer-Verlag; 2018. p. 577

[5] The Pierre Auger Collaboration et al. Observation of a large-scale anisotropy in the arrival directions of cosmic rays above 8×10^{18}eV. Science. 2017;**357**: 1266-1270. DOI: 10.1126/science.aan4338

[6] sinexx.de. Kosmische Strahlung Ist (auch) Extragalaktisch [Internet]; 2017. Available from: https://www.scinexx.de/news/kosmos/kosmische-strahlung-ist-auch-extragalaktisch/ [Accessed: August 26, 2024]

[7] sinexx.de. Rätselhafte Kosmische Strahlen Treffen Die Erde [Internet]; 2008. Available from: https://www.scinexx.de/news/geowissen/raetselhafte-kosmische-strahlen-treffen-die-erde/2008 [Accessed: August 26, 2024]

[8] Thompson DJ, Fichtel CE. Extragalactic Gamma Radiation: Use of Galaxy Counts as a Galactic Tracer. No. NASA-TM-83901; 1982

[9] Mishra RK. The universe: Search for neutrino on earth. British Journal of Multidisciplinary and Advanced Studies. 2022;**4**(1):33-62. DOI: 10.37745/bjmas.2022.0114

[10] Potter L. Telescope Array Detects Second Highest-Energy Cosmic Ray Ever [Internet]; 2023, University of Utah. Available from: https://attheu.utah.edu/facultystaff/cosmic-ray-2023/ [Accessed: August 26, 2024]

[11] Atkinson W. Single event effects from ions produced in nuclear reactions of transistors with cosmic rays. Bulletin of the American Physical Society. 2015;**18**:60

[12] Bernlöhr K. Cosmic-Ray Air Showers. Max-Plank-Institut für Kernphysik, München. Available from: https://www.mpi-hd.mpg.de/hfm/CosmicRay/Showers.html [Accessed: August 26, 2024]

[13] Ziegler JF. Terrestrial cosmic rays. IBM Journal of Research and Development. 1996;**40**(1):19-39. DOI: 10.1147/rd.401.0019

[14] Hong YM. Muon's Lite Time Measurement and High Energy Cosmic Rays Simulation. 2001. Chinese University of Hong Kong. Available from: https://www.phy.cuhk.edu.hk/newwebsite/internshipandjobs/sure/2001/yung_paper.html [Accessed: August 26, 2024]

[15] Stanev T. Cosmic Rays and Extensive Air Showers, CERN; 2010

[16] Sharma S. Atomic and Nuclear Physics. Chennai, Pearson Education India; 2008, p. 478

[17] sinexx.de. Hotspot kosmischer Strahlung Entdeckt [Internet]; 2014. Available from: https://www.scinexx.de/news/kosmos/hotspot-kosmischer-strahlung-entdeckt/ [Accessed: August 26, 2024]

[18] Bundesamt für Strahlenschutz (German Government Organisation for Radiation Protection). Natürliche Strahlung in Deutschland [Internet]. Available from: https://www.bfs.de/DE/themen/ion/umwelt/natuerliche-strahlung/natuerliche-strahlung_node.html [Accessed: August 26, 2024]

[19] Atri D, Melott AL. Cosmic rays and terrestrial life: A brief review. Astroparticle Physics. 2014;53:186-190

[20] University Kiel, Institute for Experimental and Applied Physics: Neutrino monitoring database. Was ist kosmische Strahlung? [Internet]. Available from: https://www.nmdb.eu/public_outreach/de/ [Accessed: August 26, 2024]

Chapter 3

New Peak on Thermoluminescence Glow Curve of X-Ray-Irradiated Single Crystal after Heat Treatment

Yohichi Kohzuki

Abstract

The information on solid-state (internal defect) can be contactlessly obtained from the phenomenon that light emission occurs from X-ray-irradiated KCl:Eu^{2+} single crystals. A new peak appears around 330 K on the thermoluminescence (TL) glow curve of the X-ray-irradiated crystals after the quenching from 473 to 773 K to room temperature. The peak is termed 330 K peak here. It is considered that the electrons released from *F*-centres (trapped electrons) lying in the dislocation segments are captured in the dislocation (linear crystal defect) zone and move along the dislocation line in the crystal. And further, the electrons combine with Eu^{3+} ions near the dislocation segments at 330 K. The variations of the TL glow curve and initial dislocation density under the influence of heat treatment confirm that the 330 K peak depends on the dislocations in the crystal.

Keywords: X-ray irradiation, plastic deformation, dislocation, thermoluminescence, heat treatment

1. Introduction

Some special materials emit light by mechanical treatment: microcracks, internal defects, and elastic or plastic deformation. Luminescence due to the mechanical action is called mechanoluminescence (ML). ML is divided into fractoluminescence (FL) and deformation luminescence (DL). The former appears at the fracture of material and the latter in the process of plastic deformation. ML has been investigated to develop a new sensor that detects deformation and full surface stress distribution of materials [1–3].

It is extensively known that luminescence occurs for X-ray-irradiated alkali halide crystals during plastic deformation [4]. Further, various ML (DL and FL) models have been reported for the irradiated alkali halides so far [5]. As a representative example, the mobile electron model is introduced in **Figure 1(i)-(iv)**. Alkali halide crystals such as KCl and NaCl have the structure that cations (e.g., K^+ or Na^+) and anions (e.g., Cl^-) are alternately arranged, and cation and anion vacancies are contained in the crystals (see **Figure 1(i)**). The positive and negative ion vacancies are low density; for instance, their concentration at the thermal equilibrium state is

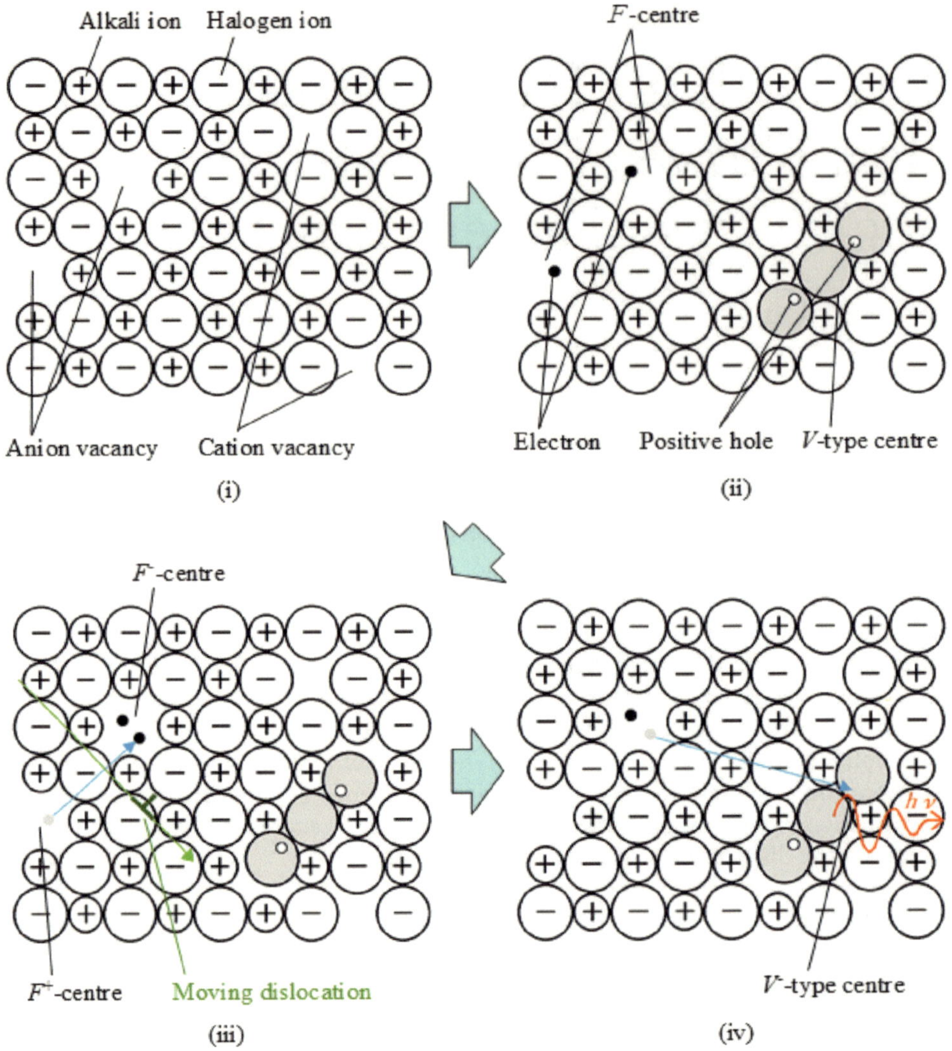

Figure 1.
An example of mobile electron model. (i) before X-ray irradiation, (ii) after X-ray irradiation (iii) under loading (dislocation movement) and (iv) recombination (luminescence).

approximately $1.2 \times 10^{-7}\%$ for the KCl single crystal quenched from 673 K to room temperature. By exposing the crystal to the X-ray irradiation, some electrons are excited and then are partly trapped at chlorine vacancies, producing *F*-centres as in **Figure 1(ii)**. Simultaneously, *V*-type centres (hole centres) are also produced in the crystal after the irradiation. These electron and hole capture centres are in a metastable state. Part of the electrons captured in these *F*-centres start to migrate along the crystal in the cause of moving dislocations (linear crystal defects) during plastic deformation or fracture (**Figure 1(iii)**). As shown in **Figure 1(iv)**, the recombination of free electrons close to an anion vacancy with *V*-type centres contributes to the thermally stimulated luminescence (TSL) then. Under a certain temperature, the electrons liberated from the created centres recombine with the trapped holes. And then, the emission energy released in the recombination process results in the thermal luminescence (TL), which is represented by $h\nu$ in **Figure 1(iv)**. Electrons trapped at shallow traps are set free from the traps at low temperatures, while electrons at deep traps begin to liberate within high temperatures. TL of KCl crystal is caused by the recombination of electrons with V_K (self-trapped hole)- or V_F (V_K-centre near a cation vacancy)-centres near 230–240 K [6].

KCl:Eu^{2+} has an excellent sensitivity as a X-ray imaging sensor utilizing an optically stimulated luminescence phenomenon in contrast to pure KCl [7]. It is extensively known that the rare-earth dopants, Eu^{2+} ions, are fluorescent and cause strong luminescence. KCl:Eu^{2+} exhibits excellent storage performance and is a reusable material for radiation therapy dosimetry [8, 9]. The TL intensity of KCl:Eu^{2+} phosphor is at least three orders of magnitude larger than that of KCl undoped with Eu^{2+} under the same reading conditions when KCl:Eu^{2+} and KCl were exposed to the same X-ray radiation dose [7]. The dopants, Eu^{2+} ions, are known as fluorescence and cause strong luminescence. **Figure 2** shows the mechanism of ML for the KCl:Eu^{2+} crystal irradiated by the X-ray. The red-colored numbers (1)–(5) exhibited in **Figure 2** correspond to the following Eq. numbers (1)–(5):

$$Eu^{2+} \rightarrow Eu^{3+} + e^{-} \quad (1)$$

$$e^{-} + F^{+} \rightarrow F \quad (2)$$

$$F + D \rightarrow e^{-} + F^{+} \quad (3)$$

$$Eu^{3+} + e^{-} \rightarrow \left(Eu^{2+}\right)^{*} \quad (4)$$

$$\left(Eu^{2+}\right)^{*} \rightarrow Eu^{2+} + h\nu\left(420\,\text{nm}\right) \quad (5)$$

where e^{-} is a free electron, F^{+} is an anion vacancy, D indicates a dislocation, $(Eu^{2+})^{*}$ is the divalent europium excited state, and $h\nu$ has its usual meaning.

X-ray irradiation causes free electrons and holes in the crystal. Some holes are trapped at Eu^{2+}, producing Eu^{3+}. The situation of Eqs. (2) and (3) is as mentioned in **Figure 1(iv)**. The electrons liberated from F_z-and F-centres combine with Eu^{3+} at each

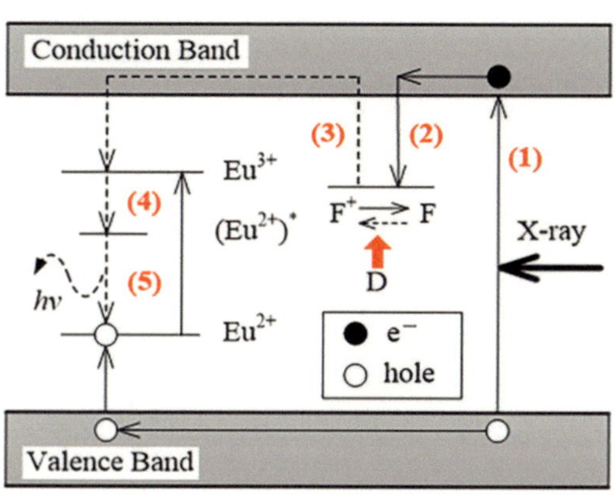

Figure 2.
Mechanism of ML (mechano-luminescence) for X-irradiated KCl:Eu^{2+}.

temperature, leading to the excited Eu^{2+}. Then, 420 nm emission occurs [10]. These processes are expressed by Eqs. (4) and (5).

Figure 3 shows the TL glow curves for KCl:Eu^{2+} (0.04 mol%) single crystals irradiated by X-ray at room temperature. These curves are concerned with various irradiation times (10 sec ~ 120 min) and were obtained by the linear heating of KCl:Eu^{2+} crystal at 24 K min^{-1}, as denoted in **Figure 4**. The defects in the crystal exposed to X-ray on the wide surface are not equally distributed in the solid. The defect concentration becomes smaller as the position of crystal is more central, since much of the energy from X-ray irradiation is consumed in forming color centres such as *F*-centres near the surface of the sample. And so, the unit of vertical scale of **Figure 3** is represented by a.u. (arbitrary unit) here. The two TL glow peaks at temperatures of 370 and 450 K are due to F_Z-and *F*-centres, respectively [12, 13]. The former peak at low temperature is hereafter termed

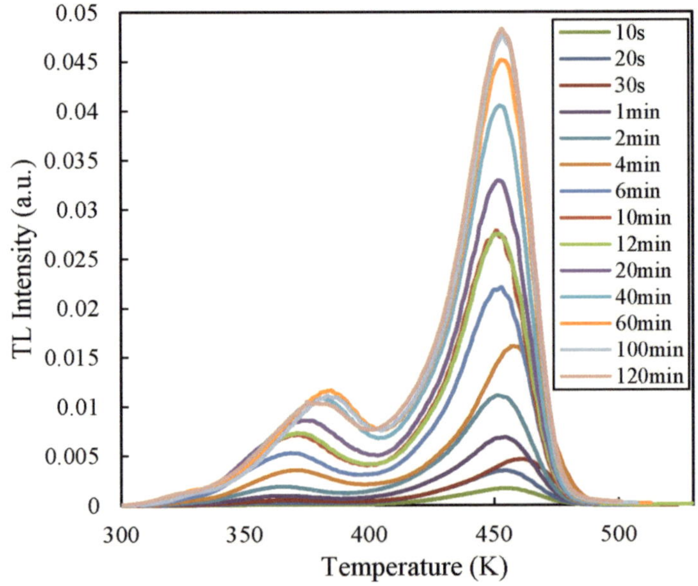

Figure 3.
Variations of TL glow curve with the irradiation time (10 sec ~ 120 min) for KCl:Eu^{2+} (0.04 mol%) single crystals irradiated by X-ray at room temperature [11].

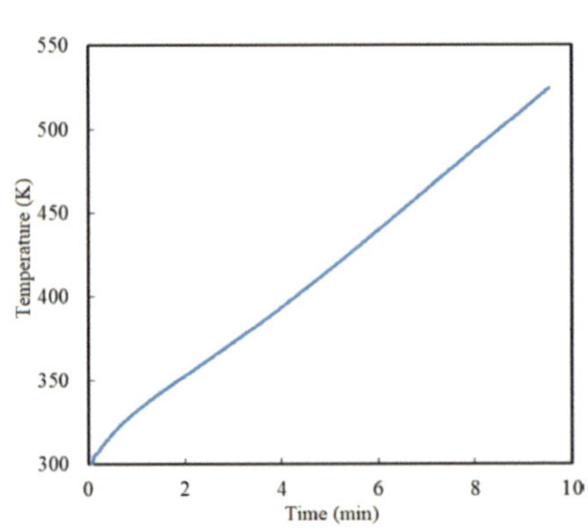

Figure 4.
Heating curve for the specimen (X-ray-irradiated KCl:Eu^{2+} crystal).

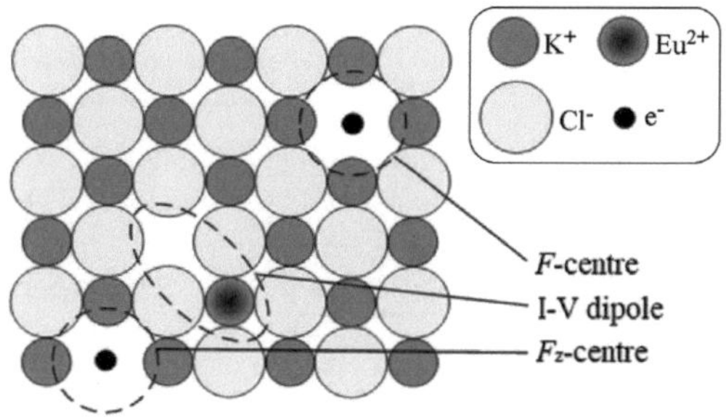

Figure 5.
Defects (F-centre, I-V dipole and F_z-centre (i.e., F-centre near I-V dipole)) in KCl:Eu^{2+}single crystal irradiated by X-ray at room temperature.

LTP, and the latter at high-temperature HTP. The electrons trapped at F_z-centres are set free from the traps at low temperature, while electrons at F-centres begin to liberate within high temperature. F_z-centres, namely seem to be shallow traps for the trapped electrons as compared with F-centres in the crystal.

The divalent cations (Eu^{2+} ions) induce cation vacancies for the purpose of conserving the electrical neutrality and are often paired with the vacancies, forming Eu^{2+} ion-vacancy dipoles for the crystals quenched from a high temperature. The pairs are termed I-V dipoles. The divalent cations strongly attract the cation vacancies in the crystal [14].

Figure 5 shows the defects in the X-ray-irradiated KCl:Eu^{2+} crystal at room temperature. By irradiating KCl:Eu^{2+} crystal with the X-ray, electron–hole pairs are made in the crystal, and chlorine vacancies store the electrons. This process produces F-centres (trapped electrons) and F_z-centres (F-centres near I-V dipoles). In **Figure 3**, the two peak heights due to F-centres and F_z-centres have become large with the X-ray irradiation time. This is because the concentration of F-centre is higher with the irradiation dose. And also, the peak temperature for F_z-centre (i.e., around 370 K of the TL glow curves) increases with its time. This may suggest that the excitonic electrons are trapped more deeply with time. While the peak for F-centre shifts to a slightly high-temperature side when the irradiation dose increases below the 6 min irradiation.

2. Experimental procedure

KCl single crystals doped with Eu^{2+} (0.03 and 0.1 mol% in melt) were prepared by cleaving the ingots, which were grown from the melt of superfine reagent powder by the vertical Bridgman method, to the size of $2.5 \times 3.5 \times 5.0$ mm^3. The samples cleaved from the ingots were thereafter heat-treated, as illustrated in **Figure 6**. Namely, the samples were annealed at 953 K for 16 h and were cooled to room temperature at 40 K/h in order to remove internal strains as much as possible. Furthermore, the samples were kept at the elevated temperature of 373 to 773 K for 7 min and then were quenched to room temperature for the purpose of dispersing the dopants (Eu^{2+}) in the crystal. Only the sample quenched from 673 K was further reannealed at 953 K for 16 h and was gradually cooled to room temperature.

The samples were subsequently exposed to the X-ray for 3 min at room temperature. The X-ray irradiation was carried out with W-target operated at 40 kV and 30 mA by Shimadzu XD-610, as shown in **Figure 7(a)**. The distance between

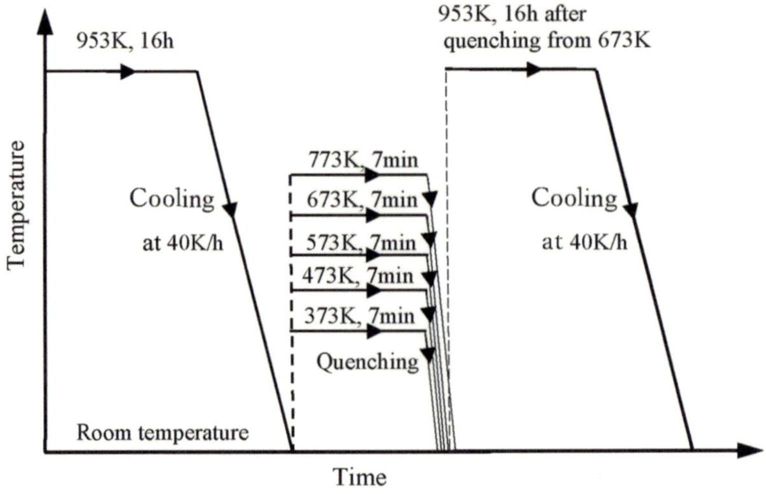

Figure 6.
Heat treatment diagram for preparing samples.

the centre of radiation source and the W-target is 0.1 m. The absorbed dose rate of KCl:Eu^{2+} at the sample position was about 5.22 Gy min^{-1}. KCl:Eu^{2+} single crystal before and after the X-ray irradiation is shown in **Figure 7(b)** and **(c)**.

The TL intensity observing system is schematically illustrated in **Figure 8**. As thermofluorescent material is heated at a constant rate, luminescence is caused at a certain temperature and its peak is observed on a TL glow curve. This was conducted by heating at the rate of 20 K min^{-1} in the dark to avoid the unwanted effect of

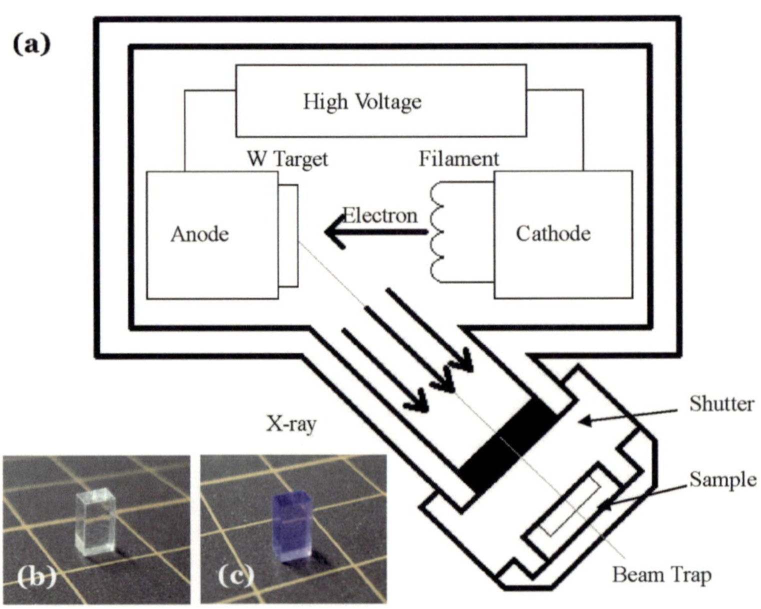

Figure 7.
(a) X-ray generation mechanism and sample (KCl:Eu^{2+} single crystal, (b) before, and (c) after the X-ray irradiation).

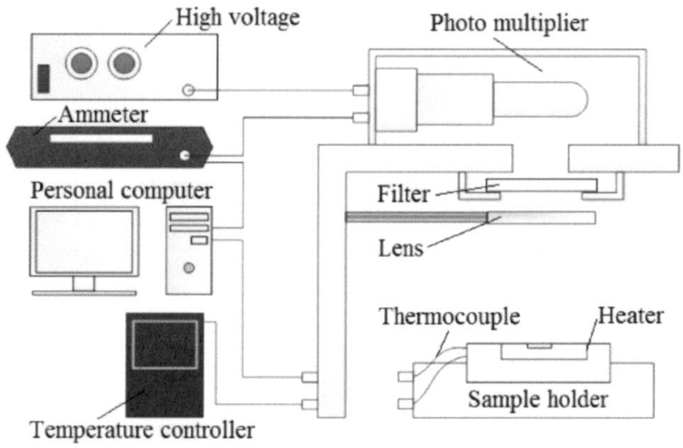

Figure 8.
Schematic diagram of TL measurement.

Experimental apparatus	Description
Temperature controller	Chino JP1131-R00
Multimeter (thermometer)	Keithley instruments DMM2000
Photo multiplier	Hamamatsu R-928
High voltage	Hamamatsu regulator DC supply: HPK-C665
Ammeter	Keithley instruments 485 picoammeter
Heater and Sample holder	Handmade

Table 1.
Experimental apparatus for TL measurements of 3 min X-ray-irradiated KCl:Eu^{2+} (0.03 mol%) single crystals after the heat treatment.

environmental light. And further, the infrared rays from the heater were excluded by using the filter. The data registration was controlled by a temperature controller (Chino JP1131-R00). TL measurements were carried out at 300–523 K using a photo multiplier (Hamamatsu R928) with a sensitive region from 185 to 900 nm through a lens. The details of the TL observing system were described elsewhere [15]. The various apparatuses for TL are given in **Table 1**. Luminescence intensity was recorded in a personal computer and then the data were processed.

The density of dislocations in some samples prepared by the above-mentioned treatment was measured by etch pits technique. Corrosive liquid was the saturated solution of $PbCl_2$ in ethyl alcohol, adding two drops of water [16]. The etching on a (100) plane of the samples was performed at room temperature for 10 min.

3. Results and discussion

3.1 TL glow curves of X-ray-irradiated KCl:Eu^{2+} crystals after various heat treatments

Figure 9(a) shows the TL glow curves of 3 min X-ray-irradiated KCl:Eu^{2+} (0.03 mol%) single crystals after the quenching from 373 to 773 K to room temperature. Scaled-up representations of **Figure 9(a)** are shown in **Figure 9(b)** within

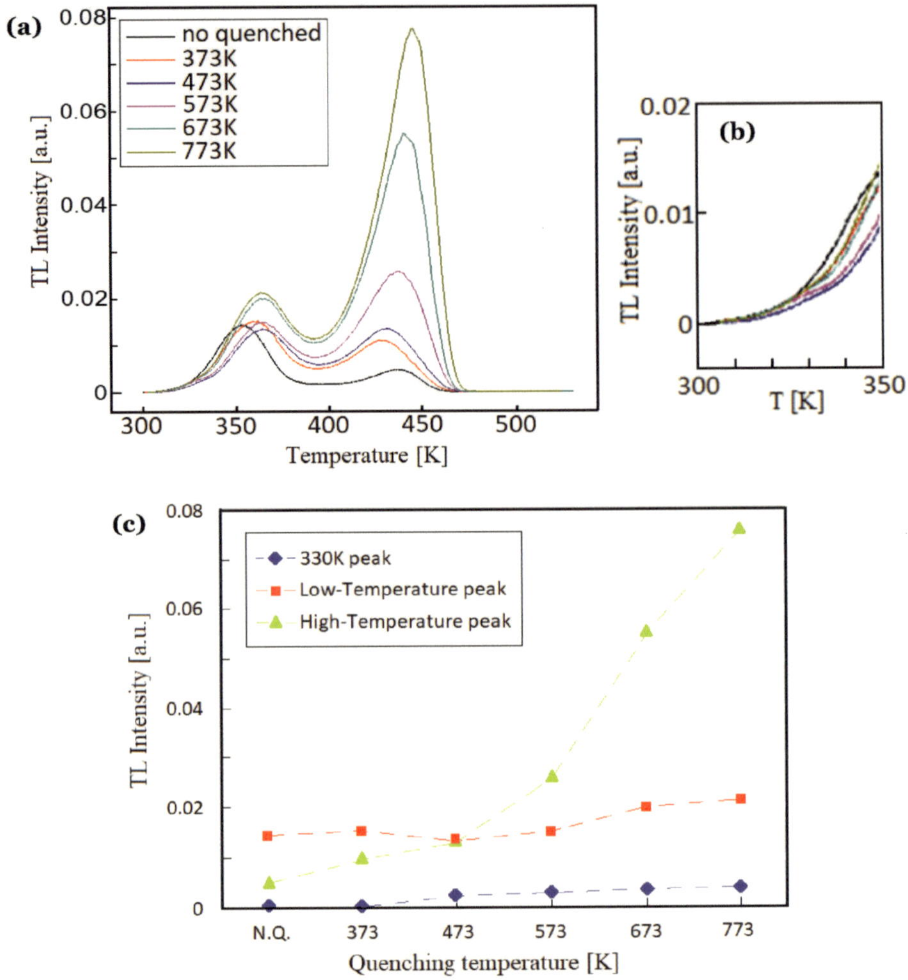

Figure 9.
(a) TL glow curves and (c) (◆) 330 K peak, (■) LTP, and (▲) HTP intensities on the TL glow curves for 3 min X-ray-irradiated KCl:Eu^{2+} *(0.03 mol%) single crystals after quenching from various temperatures T: (——) 373 K, (——) 473 K, (——) 573 K, (——) 673 K, (——) 773 K, and (——) N.Q. (no quenched) [17]. These curves in* ***Figure 9(b)*** *are scaled-up representations of the corresponding curves in* ***Figure 9(a)****.*

the temperature range of 300–350 K. The height variations of TL glow peaks in **Figure 9(a)** with the quenching temperature are shown in **Figure 9(c)** for the crystals, which are obtained taking account of base line. The data for the only irradiated sample without quenching treatment are also shown in these figures to examine the influence of quenching on the TL glow peaks.

As can be seen from the TL intensity for each peak in **Figure 9(c)**, both LTP due to F_Z- centres and HTP due to F-centres tend to be large with the quenching temperature. The new peak appears around 330 K on some TL glow curves in **Figure 9(a)** and **(b)**, which is named the 330 K peak here. Concerning the 330 K peak, the peak intensity is zero, independent of the quenching temperature below about 373 K. However, it appears on the TL glow curves for the crystals quenched from 473 K and is barely larger with the quenching temperature.

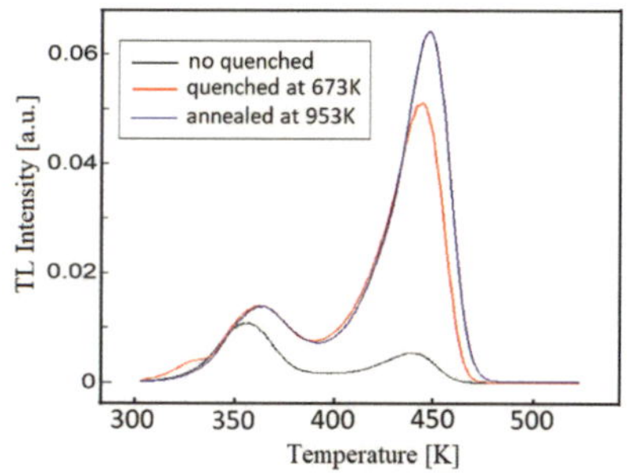

Figure 10.
Variant TL glow curve of 3 min X-ray-irradiated KCl:Eu^{2+} (0.03 mol%) single crystals after the following treatments: (——) no quenched; (—) quenching from 673 K; (—) quenching from 673 K → annealing at 953 K [17].

The black solid line in **Figure 10** is the TL glow curve for the only irradiated crystal. While, the red solid one in **Figure 10** corresponds to the green-colored one in **Figure 9(a)**, that is to say, for 3 min X-ray-irradiated KCl:Eu^{2+} (0.03 mol% in the melt) single crystal after the quenching from 673 K to room temperature. With regard to the irradiated crystal previously quenched from 673 K and subsequently annealed at 953 K, it is represented by the blue solid line in **Figure 10**.

As observed in **Figure 10**, the 330 K peak becomes to appear around 330 K on the red-colored TL glow curve by quenching the KCl:Eu^{2+} crystal from 673 K. Annealing the crystal after the quenching heat treatment, the 330 K peak disappears on the TL glow curve (blue solid line).

The energy level diagram of the 330 K peak for the crystal is illustrated in **Figure 11**, where the new centre is considered to be the *F*-centre nearby the dislocation segment. And further, the new centre has a similar TL mechanism as the *F*-centre, as denoted by Eqs. (4) and (5). That is to say, the electrons released from the *F*-centres lying in the expansion region of the dislocation segments are captured in the dislocation zone. The dislocation-captured electrons move along the dislocation line under Coulomb attraction caused by the Eu^{3+} ions and combine with Eu^{3+} ions near the dislocation segments at 330 K, leading to the (Eu^{2+})*. Then, (Eu^{2+})* transforms to the ground state, emitting the characteristic light of wavelength 420 nm.

From this model, it may be deduced that the 330 K peak intensity is influenced by the dislocation density in the crystal. This is described in the next section (i.e., Section 3.2).

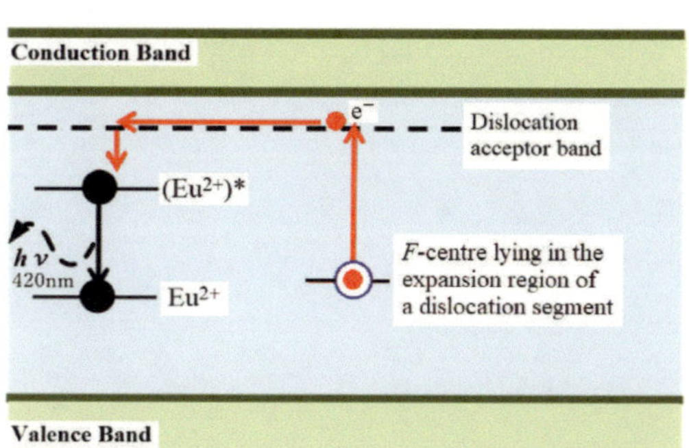

Figure 11.
Schematic energy level diagram of 330 K peak for X-ray-irradiated KCl:Eu^{2+} crystal after quenching treatment from the temperatures 473–773 K (reproduced from Figure 9 in Ref. [18]).

3.2 330 K peak intensity dependent on the density of dislocations in the crystal

Initial dislocation density dependence of quenching temperature is shown as solid circles in **Figure 12(a)** for KCl:Eu^{2+} (0.1 mol%) single crystals. Etch pits technique was used to detect the density of dislocations. The density of dislocations in the crystal is almost constant (0.83×10^5 cm^{-2}) independently of quenching temperature below about 400 K. Above the temperature, it becomes larger with the quenching temperature. Especially, the increment is remarkable at 573–673 K. Such a tendency of the dislocation density is similar to the variation of 330 K peak intensity with the quenching temperature at 300–773 K, as can be seen from the blue dotted line in **Figure 9(c)**. Namely, this experimental result on the variation of dislocation density supports that the appearance of 330 K peak on the TL glow curve is caused by the phenomenon based on the dislocation segments in the crystal.

Figure 12(b) and **(c)** are the optical micrographs of the etch pits for KCl:Eu^{2+} (0.1 mol%) quenched from 673 K and for annealed at 953 K for 16 h after the quenching from 673 K, respectively. The etched planes at plots represented by "micrograph (b)" and "micrograph (c)" in **Figure 12(a)** are displayed in **Figure 12(b)** and **(c)**. Although it is difficult to resolve the individual etch pits for high dislocation density, the measurement of dislocation density in a crystal has been carried out by the etch pit technique. The position of dislocation after the heat treatments is marked by a pyramidal pit, while the position where a dislocation slipped out on a slip plane in the crystal after the treatment is marked by a flat-bottom pit. The dislocation densities on a (100) plane are found to be 5.18×10^5 cm^{-2} for the crystal quenched from 673 K and 1.23×10^5 cm^{-2} for the crystal quenched from 673 K and subsequently annealed at 953 K for 16 h. The dislocation density decreases by annealing the crystal quenched from 673 K. This remarkable decrease in dislocation density is considered to cause the disappearance of the 330 K peak on the TL glow curve (blue solid line) observed in **Figure 10**. This also supports the idea that the 330 K peak intensity depends on the dislocation density of the crystal.

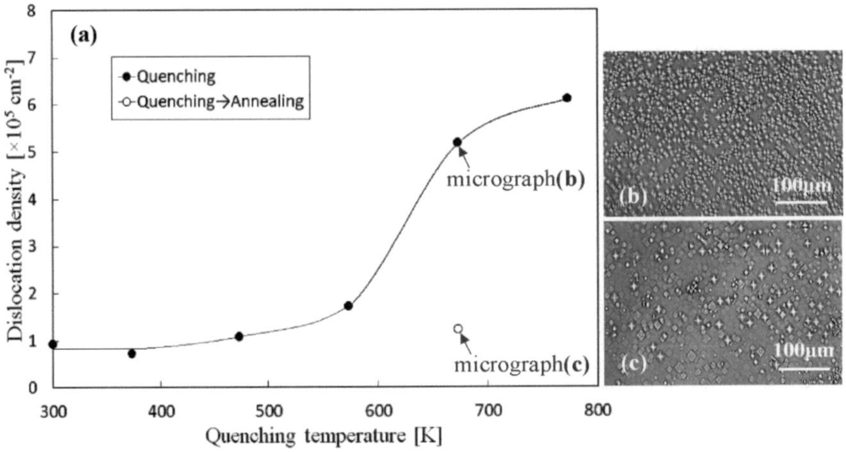

Figure 12.
(a) Initial dislocation density on (100) planes of KCl:Eu^{2+} (0.1 mol%) single crystals quenched from (●) each temperature from 300 to 773 K and prepared by (○) the heat treatment: annealing at 953 K after quenching from 673 K (reproduced from Ref. [17] with permission from the publisher). Optical micrographs of the etch pits for the samples, (b) quenched from 673 K, and (c) quenched from 673 K and subsequently annealed at 953 K.

DOI: http://dx.doi.org/10.5772/intechopen.1007612

4. Conclusion

The new peak (i.e., 330 K peak) is found to appear on the TL glow curves of 3 min X-ray-irradiated $KCl:Eu^{2+}$ (0.03 mol%) single crystals after the quenching from 473 to 773 K to room temperature. However, by X-ray-irradiating the annealed $KCl:Eu^{2+}$ crystal at 953 K after quenching treatment from 673 K, the 330 K peak disappears on the TL glow curve. This annealing treatment causes a remarkable decrease in dislocation density. This leads to the phenomenon that the 330 K peak intensity is variable depending on the density of dislocations in the crystal. This result will provide us with a useful method to investigate the deformation structure concerned with dislocations in materials contactless and will be applied to new stress sensors in the future.

Acknowledgements

The author is pleased to give thanks to Dr. T. Ohgaku for helpful and constructive discussions on the chapter and to S. Matsuura and Y. Igari for their experimental assistance.

Conflict of interest

The author declares no conflict of interest.

Author details

Yohichi Kohzuki
Department of Mechanical Engineering, Saitama Institute of Technology, Fukaya, Japan

*Address all correspondence to: kohzuki@sit.ac.jp

References

[1] Ishihara T, Tanaka K, Hirao K, Soga N. Fracto-luminescence of rare earth element-doped hexacelsian ($BaAl_2Si_2O_8$). Japanese Journal of Applied Physics. 1997;**36**:L781-L783. DOI: 10.1143/JJAP.36.L781

[2] Kim JS, Kwon YN, Shin N, Sohn KS. Visualization of fractures in alumina ceramics by mechanoluminescence. Acta Materialia. 2005;**53**:4337-4343. DOI: 10.1016/j.actamat.2005.05.032

[3] Jia Y, Yei M, Jia W. Stress-induced mechanoluminescence in $SrAl_2O_4:Eu^{2+},Dy^{3+}$. Optical Materials. 2006;**28**:974-979. DOI: 10.1016/j.optmat.2005.05.014

[4] Zakrevskii VA, Shuldiner AV. Electron emission and luminescence owing to plastic deformation of ionic crystals. Philosophical Magazine. B. 1995;**71**:127-138. DOI: 10.1080/01418639508240301

[5] Chandra BP. In: Vij DR, editor. Luminescence of Solids. New York: Springer; 1998. pp. 309-314

[6] Egemberdiev Zh, Elango A, Zazubovich S. Luminescence accompanying the recombination of VF and F centres in KCl and KBr crystals. Physica Status Solidi (b). 1980;**97**:449-456. DOI: 10.1002/pssb.2220970208

[7] Camacho AQ, Muñoz GH, Rubio JO, Garcia JM, Murrieta HS, Hernandez JA. Dosimetric properties of KCI:Eu. Journal of Materials Science Letters. 1988;7:437-440. DOI: 10.1007/BF01730681

[8] Driewer JP, Chen H, Osvet A, Low DA, Li H. Radiation hardness of the storage phosphor europium doped potassium chloride for radiation therapy dosimetry. Medical Physics. 2011;**38**:4681-4688. DOI: 10.1118/1.3611043

[9] Hansel RA, Xiao Z, Zhang L, Li HH. X-ray storage performance of $KCl:Eu^{2+}$ with high cumulated dose. Nuclear Instruments and Methods in Physics Research Section B: Beam Interactions with Materials and Atoms. 2014;**326**:178-181. DOI: 10.1016/j.nimb.2013.08.067

[10] Rubio JO, Flores MC, Murrieta HS, Hernandez JA, Jaque F, Lopez FJ. Influence of concentration and aggregation-precipitation state of divalent europium in the room-temperature coloring of KCl. Physical Review B. 1982;**26**(4):2199-2207. DOI: 10.1103/PhysRevB.26.2199

[11] Kohzuki Y. Luminescence of X-ray-irradiated single crystals under the influence of compression. In: Maghraby AM, editor. Luminescence – Emerging New Applications. London, UK, London: IntechOpen; 2024. DOI: 10.5772/intechopen.1005364

[12] Aceves R, Pérez Salas R, Barboza-Flores M. The role of F centres in the thermoluminescence of low-energy uv- and x-irradiated $KCl:Eu^{2+}$. Journal of Physics: Condensed Matter. 1994;**6**:10397-10405. DOI: 10.1088/0953-8984/6/47/022

[13] Pedroza-Montero M, Castañteda B, Meléndrez R, Chernov V, Barboza-Flores M. Comparative investigations of TL and OSL in $KCI:Eu^{2+}$ crystals irradiated with uv and X-rays. Radiation Effects and Defects in Solids. 2001;**154**:319-324. DOI: 10.1080/10420150108214071

[14] Pick H, Weber H. Dichteänderung von KCl-Kristallen durch Einbau zweiwertiger Ionen. Zeitschrift für

Physik. 1950;**128**:409-413. DOI: 10.1007/BF01339441

[15] Kohzuki Y, Ohgaku T. Study on luminescence of KCl:Eu^{2+} crystals after X-ray irradiation at room temperature. Crystals. 2019;**9**(7):331. DOI: 10.3390/cryst9070331

[16] Kataoka T. Studies on Plastic Deformation of Alkali Halide Crystals. Doctoral Thesis,. Osaka: Osaka University; 1972. pp. 30-34

[17] Kohzuki Y. Influence of compressive load and heat treatment on deformation luminescence of KCl:Eu^{2+} single crystals exposed to X-ray. Radiation Effects and Defects in Solids. 2024;**179**:605-620. DOI: 10.1080/10420150.2023.2294467

[18] Kohzuki Y, Ohgaku T. Deformation luminescence of X-ray-irradiated KCl:Eu^{2+} single crystals by compression. Journal of Luminescence. 2023;**253**:119469. DOI: 10.1016/j.jlumin.2022.119469

Chapter 4

Advancements of Nanomaterials for Diagnostic Imaging, Radiotherapy, and Nuclear Medicine

Wubshet Getachew Mengesha, Cherie Sisay Mekonnen and Ambaye Fisehaw Tesfaw

Abstract

Among the main purposes in the medical industry for prevention, diagnosis, and treatment of diseases are controlling radiation doze using a state-of-the-art medical technology, and optimizing image quality. The study of radiation properties and their effect and interaction, medical imaging analysis, and nuclear medicine offer unwavering interest. Nanomaterials play a key role in the innovation, design, and development of medical technologies in health care by testing, implementation, and, development of new diagnostic and medical equipment. Because of their unique characteristics and properties, they have become the most chooseable and multi-purpose devices in the medical industry. They have grasped and gained considerable attention and offer unwavering interest due to their wide extent of technological applications, especially in the medical sector. Nanomaterials have been playing a significant role in the medical sector in the production of medical devices and bone and joint replacement biosensors for replacing damaged organs, correcting abnormalities, and thus improving the quality of life of patients. In this paper, we have attempted to highlight the applications and important progress achieved by nanomaterials to give a comprehensive understanding of various medical techniques with radiation principles, medical imaging, radiotherapy, and nuclear medicine. Finally, the challenges and future trends for the development and design of safe nanomaterial for medical technology and applications have been suggested.

Keywords: nanomaterials, radiation, diagnostic imaging, radiotherapy, nuclear medicine, radioisotopes, radiopharmaceuticals

1. Introduction

The application of nanomaterials in the medical industry is currently an active field because of old materials and properties from bulk scale modified with new ones into the nanoscale. Materials play a privileged role since they often lead to new unexpected phenomena, such as superconductivity, magnetic effects, or new topologies [1–3]. These new overall properties, nanomaterials offer exciting opportunities in medicine for the detection and treatment of diseases due to their unique characteristics and properties to enhance drug targeting capacities, bioavailability, and biocompatibility [4–6].

Material-based technologies have been becoming a fundamental part of our daily lives, starting from the development of electronic circuits, space and satellite technologies, and the improvement of medical technologies. We are observing that for tackling the big issues in the world today, developing new technologies to solve the most challenging problems like combating climate change, adopting drug discovery mechanisms to fight disease, ensuring data security issues, addressing energy needs and storage technologies, and so on to improve our lives [7–9].

Nanomaterials, with their unique properties at the nanoscale level, have revolutionized the medical field. The continuous progression of medical technologies has significantly improved the quality, accessibility, and personalization of healthcare services by enhancing patient outcomes and transforming the way medicine is delivered, enabling pioneering solutions that address unmet medical needs and effective medical interventions. Modern clinical practice and research are heavily dependent on technology, and hospitals are equipped with nanomaterial devices for clinical measurement, diagnosis, and treatment [5–11]. Nanomaterials possess unique physical, chemical, biological, thermal, radiation, and magnetic properties that make them highly versatile for applications in diagnostic, imaging, therapy, and nuclear medicine. Continued research and development in nanotechnology hold promise for advancing precision medicine and improving clinical outcomes for patients with cancer and other diseases [12–17].

Radiation technology has attracted a much broader base of professionals and is a rapidly growing field that provides wide applications in medicine for radiation protection, radiation therapy, diagnostic, imaging, treatment of human disease, radiation oncology, radiation safety, and radiation dosimeter programs [6, 18]. For safety and quality assurance in the use of radiation for medical purposes targeted to reduce unnecessary radiation risks, nanomaterial has played a pivotal role in the diagnostic imaging and treatment of disease with ionizing radiation and provided the scientific base, initially for the understanding of the production of radiation (its interaction with matter, measurement parameters, and dose) for the technological development of medical equipment [9, 10, 19, 20].

Figure 1 illustrates the central role of nanomaterials in medicine, highlighting their applications in diagnostic imaging, radiotherapy, and nuclear medicine (radiopharmaceuticals and radioisotopes). Each of these fields is interconnected, showcasing the interdisciplinary impact of nanomaterials in advancing medical technologies.

In the following basic subsections, some of the most impactful and innovative applications have been described among the groundbreaking advancements that have revolutionized the medical industry for diagnosis imaging, therapy, and nuclear medicine, as described in **Figure 1**. In the first part, diagnostic imaging principles are described. Then, the development of medical technologies like imaging and laser, radioisotopes, and magnetic materials used for the medical industry has been listed in their subtitles. In the final section the challenges and future trends, with the conclusion, for the development and design of safe nanomaterial for medical technology has been suggested.

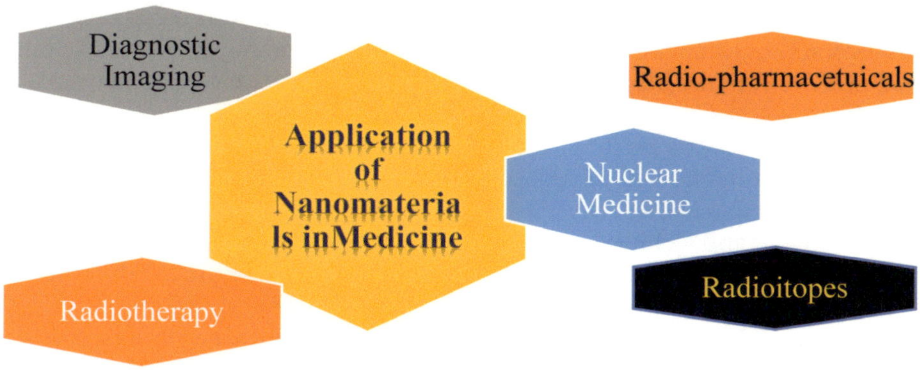

Figure 1.
Application of nanomaterials in medicine (diagnostic imaging, radiotherapy, and nuclear medicine).

2. Diagnostic imaging

Medical image analysis for disease detection and diagnosis is a rapidly evolving field that holds immense potential for improving healthcare outcomes [2, 5]. Diagnostic imaging involves understanding the physical principles underlying various imaging modalities and technologies used to optimize image quality, ensure patient safety, and improve diagnostic accuracy which enables the study of the function and morphology of the human body in detail [21–25]. However, modern medical imaging includes not only image production but also image processing, computer-aided diagnosis, artificial intelligence-assisted, image recording and storage, image transmission, and many more. There have been remarkable improvements in medical imaging, diagnostic, and therapeutic procedures toward human health in various modalities [26–28]. For example, superparamagnetic iron oxide nanoparticles provide for MRI contrast, gold nanoparticles for x-ray attenuation for CT scans, and radioisotope nanoparticles for visualizing metabolic processes. Ensuring minimal radiation exposure to patients while maintaining diagnostic image quality is a primary concern [3, 29, 30]. The medical industry, in general, and the medical imaging modality, in particular, have considerably advanced through different modalities and techniques. Among these accomplishments, nanomaterials played a significant role and took a lion's share [1, 31, 32].

Ensuring minimal radiation exposure to patients while maintaining diagnostic image quality is a primary concern, particularly with modalities like X-ray and CT. Balancing effective dose and image quality requires continual vigilance and optimization. Achieving high-resolution images that accurately represent the patient's condition is complex. It involves adjusting various parameters and technologies to enhance diagnostic utility while avoiding artifacts and distortions [11, 30, 33]. Keeping pace with rapid advancements in imaging technology and integrating new modalities and techniques into clinical practice pose ongoing challenges. This includes adapting to new hardware, software, and imaging protocols [34].

Regular calibration and maintenance of imaging equipment are crucial to ensure reliable and consistent performance which requires systematic quality assurance programs and technical expertise to identify and rectify any issues. Adhering to regulatory standards and guidelines for radiation safety, equipment performance, and imaging protocols requires ongoing education and vigilance. Addressing these challenges involves continuous education, technological innovation, and interdisciplinary collaboration to enhance diagnostic imaging practices and improve patient outcomes [1, 25, 35].

Research continues on methods to minimize radiation exposure without compromising image quality. This includes advancements in imaging technology, such as improved detectors and advanced image reconstruction algorithms. Tailoring imaging protocols and techniques to individual patient needs based on their unique anatomical and physiological characteristics could lead to more precise diagnoses and better patient outcomes [28, 36].

Using different imaging technologies and techniques to provide a more comprehensive view of patient conditions and to improve image reconstruction algorithms to reduce noise, enhance resolution, and speed up imaging processes. Improving data management systems to handle the increasing volume of imaging data, ensuring secure storage, retrieval, and sharing of images while maintaining patient privacy [22, 37]. These research areas aim to push the boundaries of diagnostic imaging, enhancing both the quality of patient care and the efficiency of diagnostic processes [25, 26].

3. Radiotherapy

Radiation technology has attracted a much broader base of professionals and is a rapidly growing field that provides a wide application in medicine in radiation protection, radiation therapy, diagnostic imaging, disease treatment, radiation oncology, and radiation safety and radiation dosimeter programs [38, 39]. The use of ionizing radiation has changed enormously for the treatment of cancer since the discovery of X-rays and radium. These initial efforts stimulated a revolution of conceptual and technological innovations that formed the basis of the safe and effective therapies used today [40–43].

Radiotherapy is a medical treatment that uses high-energy radiation to destroy cancer cells or shrink tumors and damage the DNA within the cancer cells, inhibiting their ability to grow and divide, ultimately leading to cell death [33, 44, 45]. It can be administered externally using machines or internally through the placement of radioactive sources near or within the tumor as well as using advanced radiotherapy techniques to enhance the precision and effectiveness of treatment, offering better outcomes and fewer side effects for patients [2, 35, 45, 46].

Radiotherapy can cause side effects in healthy tissues near the treatment area. Managing these side effects and minimizing their impact on the patient's quality of life is a key challenge [1, 39, 44, 47, 48]. Adapting radiotherapy to individual patients based on their specific disease type, genetics, and overall health can be complex and requires ongoing research and adaptation of treatment strategies [33, 39]. Addressing these challenges involves continued research, technological innovation, and a multidisciplinary approach to enhance the effectiveness and safety of radiotherapy for patients [49–51].

Studying the long-term effects of radiotherapy, including late-onset side effects and quality-of-life issues, helps in developing strategies for follow-up care and improving survivorship outcomes. Research advancements in radiotherapy equipment, such as more precise treatment machines or improved quality assurance technologies, contribute to better treatment delivery and safety. These research areas collectively aim to advance the field of radiotherapy, making treatments more effective, precise, and patient-friendly while addressing the challenges faced in current practices [38, 40, 50, 52].

Therapeutic radiopharmaceuticals are used to treat a previously diagnosed disease or disorder. Radiopharmaceutical therapy is emerging as a safe and effective targeted approach to treating many types of cancer [53].

4. Nuclear medicine

Nuclear medicine is a highly multidisciplinary specialty that develops and uses instrumentation and radiopharmaceuticals to study physiological processes for diagnosing and treating diseases. It is most often used to evaluate the function of any organ or structure in the body, diagnose and stage various diseases, and treat certain illnesses. It involves the use of radioactive material to diagnose and treat different conditions by providing both structural and functional analysis of conditions such as certain types of cancer and neurological and heart diseases [20, 54].

The medical technologies that are used for the treatment and diagnostic applications have started with the invention of X-rays and radioactivity, which could open a new era of nuclear medicine, which is the way of using radionuclides and radioisotopes. Due to the identification of diseases and treatment processes being complex and slow, advances in the medical sector have been showing inspired progress to facilitate, simplify, and reduce the costs of delivering improved health care [3, 20, 21]. Nuclear medicine uses radioactive tracers (radiopharmaceuticals) inside the body to see how organs are functioning (for diagnosis) or to destroy damaged organs (for treatment). It is expanding rapidly, especially due to new imaging technologies that improve the accuracy of detection, localization, and characterization of diseases [55, 56]. Current clinical applications of nuclear medicine include the ability to diagnose diseases, assess therapeutic response by reducing exposure with effective treatment methods, and develop new cost-effective and state-of-the-art medical technologies (higher resolution, faster, and more accurate imaging instruments) to improve disease diagnosis and treatment.

In the next subsections, the role of radioisotopes and radiopharmaceuticals in nuclear medicine, both for diagnosis and therapeutic purposes, is highlighted.

4.1 Radioisotopes

From the beginning of discoveries of atoms, from the ideas of the Greek philosophers, goes to the discovery of natural radioactivity at the end of the nineteenth century, then the understanding of the structure of atoms, the discovery of elemental isotopes, and finally the discovery of radioactive decay have played the critical roles in the functioning of nuclear medicine and technology [40, 41, 57, 58].

In nature, many radioisotopes have medical applications that have been developed by nuclear science and technology. Radioactive tracers have been widely used for diagnostic purposes and the treatment of diseases. Such use is generally known as nuclear medicine. Radioisotopes have applications in diagnosis, therapy, and health care including radiopharmaceuticals [5, 7, 13]. The medical industry has been greatly influenced by the development of radioisotopes for imaging techniques, dosimeters, and radiation protection. Recently, artificially produced radioactive isotopes used in medicine have taken from the immediate discovery of natural and later artificial radioisotopes for several medical applications [9, 39, 59].

So, for the evolution of new technologies in the medical industry, a radioisotope-based technology has led to improved diagnostic capabilities, more effective treatments, and enhanced patient safety.

4.2 Radiopharmaceuticals

Radiopharmaceuticals are substances that contain a radioisotope and have properties that make them effective markers in medical diagnostic or therapeutic procedures. Radiopharmaceuticals contain small amounts of radioisotopes that can be produced by irradiating a specific target inside particle accelerators under carefully controlled conditions and validated standard operating procedures [30, 34, 47, 60]. These radiopharmaceuticals for imaging organ function and disease involvement are the unique capabilities of nuclear medicine. Radiopharmaceuticals used in diagnostic nuclear medicine procedures generally emit either gamma radiation or positrons, and the half-lives of radionuclides for imaging applications generally span from minutes to several hours [54, 60, 61].

Radiopharmaceuticals are used in small amounts to evaluate the function of different organs. For medical treatment and diagnostic purposes, a radioisotope (composed of either a standalone radionuclide or an organic molecule) is attached and delivered by inhalation, oral, intravenous, or interstitial route used for selected purposes to specific organs, tissues, or cells. The images captured with the detector from the radiation emitted by the radiopharmaceuticals will be analyzed [57, 58, 60, 61].

5. Future developments in nanomaterial for medical applications

The field of nanomaterials is rapidly evolving, promising groundbreaking advancements in medical imaging, therapy, and nuclear medicine [2, 22, 23]. While nanomaterials offer immense potential, challenges such as biocompatibility, toxicity, and long-term effects need to be carefully addressed. Future research will focus on developing more sophisticated nanomaterials with improved properties and targeted delivery systems [27, 41, 45, 50]. Technical progress has produced a variety of novel imaging techniques and has enormously extended diagnostic opportunities. Advanced technical improvements, introduction of new imaging modalities, and progressive use of computer technology in diagnostics (like image generation and interpretation, processing and archiving, and so on) should be considered, for the improvement of current imaging procedures [62–64].

In the time and age of big data, healthcare systems have been shifting from the traditional focus on safety and regulatory compliance to new artificial intelligence for supporting the information-based use of technology and accurate data analytics. A key issue especially faced in imaging is the design of an imaging probe which includes biocompatible, easily excreted from the body, signal, and sensitivity, targeting to the area of interest, the selection of a radioisotope with sufficient half-life and optimal energy, low toxicity, and good bio-distribution. Nanomaterials can potentially overcome these limitations due to their advantageous size and physiochemical properties [2, 65, 66].

Figure 2 has outlined focusing on nanomaterials for medical applications with in three areas. Future challenges include biocompatibility, toxicity, and long-term effects. Secondly, on the research focus, on developing sophisticated nanomaterials, improving imaging techniques, and targeted delivery. Thirdly, the key issues involve

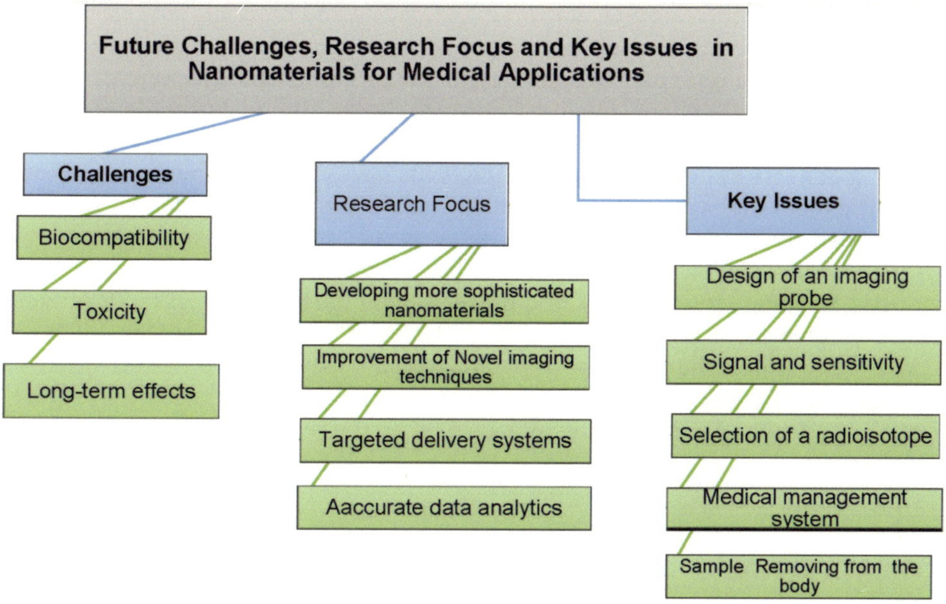

Figure 2.
Future developments in nanomaterial for medical applications.

the design of imaging probes, signal sensitivity, and the selection of radioisotopes, as well as the removal of samples from the body. This provides a comprehensive overview of the critical aspects in nanomedicine.

In recent years there have been so many extraordinary improvements that have been made toward the development of medical technologies, diagnosis, therapeutic, and treatment of diseases. However, with its increasing incidence, medical management continues to be a challenge for the future.

6. Conclusion

The application of nanomaterials in the medical field has transformed diagnostic imaging, radiotherapy, and nuclear medicine, offering unprecedented opportunities for improved patient care and outcomes. They possess unique physical, chemical, biological, thermal, radiation, and magnetic properties that have been leveraged to enhance the capabilities of various medical technologies. In diagnostic imaging, nanomaterials have played a pivotal role in improving image quality, contrast, and safety. Superparamagnetic iron oxide nanoparticles, for instance, have enhanced MRI contrast, while gold nanoparticles have improved X-ray attenuation for CT scans. Continuous research aims to further optimize imaging techniques, minimize radiation exposure, and develop personalized imaging protocols tailored to individual patient needs. Advancements in data management, image processing, and artificial intelligence-assisted diagnostics have also contributed to the evolution of modern medical imaging.

For therapeutic applications, nanomaterials have changed treatment approaches, leading to more targeted and effective radiation delivery. Nanoparticle-assisted radiation therapy and other nanotechnology-enabled techniques have expanded the options for cancer treatment, improving targeting and reducing side effects. Moreover, nanomaterials have enhanced radiation protection and dosimetry, ensuring greater safety for both patients and healthcare professionals. The application of nanomaterials in nuclear medicine has improved the specificity, bioavailability, and stability of radioisotopes, leading to more effective radiopharmaceuticals and targeted drug delivery systems. These advancements have enhanced the diagnostic and therapeutic capabilities of nuclear medicine, contributing to earlier disease detection and more personalized treatment strategies.

As the medical field continues to evolve, the integration of nanomaterials into diagnostic, imaging, radiotherapy, and nuclear medicine technologies holds immense promise. Ongoing multidisciplinary collaboration and continued research will be crucial to fully realize the potential of nanomaterial in advancing precision medicine, improving patient outcomes, and addressing unmet clinical needs. The future of healthcare is poised to be significantly shaped by the remarkable capabilities of nanomaterials.

Conflict of interest

The authors declare that there is no conflict of interest, and the paper was conducted in the absence of any financial interests or personal relationships that could be understood as a potential conflict of interest.

Author details

Wubshet Getachew Mengesha*, Cherie Sisay Mekonnen and Ambaye Fisehaw Tesfaw
Department of Physics, College of Natural and Computational Sciences, Woldia University, Woldia, Ethiopia

*Address all correspondence to: wubshet2003@gmail.com

References

[1] Siddique S, Chow JCL. Application of nanomaterials in biomedical imaging and cancer therapy. Nanomaterials. 2020;**10**(9):1-41. DOI: 10.3390/nano10091700

[2] Phua VJX et al. Nanomaterial probes for nuclear imaging. Nanomaterials. 2022;**12**(4):1-23. DOI: 10.3390/nano12040582

[3] Gul S, Khan SB, Rehman IU, Khan MA, Khan MI. A comprehensive review of magnetic Nanomaterials modern day Theranostics. Frontiers in Materials. 2019;**6**(July):1-15. DOI: 10.3389/fmats.2019.00179

[4] Martin FL, Pollock HM. Microspectroscopy as a tool to discriminate nano-molecular cellular alterations in biomedical research. Oxford Handbook of Nanoscience and Technology. 2009;**2**(October):285-336. Available from: http://eprints.lancs.ac.uk/26720/

[5] Pellico J, Gawne PJ, De Rosales RTM. Radiolabelling of nanomaterials for medical imaging and therapy. Chemical Society Reviews. 2021;**50**(5):3355-3423. DOI: 10.1039/d0cs00384k

[6] Su S, Kang PM. Systemic review of biodegradable nanomaterials in nanomedicine. Nanomaterials. 2020;**10**(4):1-21. DOI: 10.3390/nano10040656

[7] Beyer T, Bailey DL, Birk UJ, Buvat I, Catana C, Cheng Z, et al. Medical Physics and imaging-a timely perspective. Frontiers in Physics. 2021;**9**:1-12. DOI: 10.3389/fphy.2021.634693

[8] Malik S, Muhammad K, Waheed Y. Emerging applications of nanotechnology in healthcare and medicine. Multidisciplinary Digital Publishing Institute (MDPI). 2023;**28**(18):6624. DOI: 10.3390/molecules28186624

[9] Goel M, Mackeyev Y, Krishnan S. Radiolabeled nanomaterial for cancer diagnostics and therapeutics: Principles and concepts. Cancer Nanotechnology. 2023;**14**(1):1-36. DOI: 10.1186/s12645-023-00165-y

[10] Agarwal AS. Applications of mathematics and Nuclear Physics in medicine. International Journal for Research in Applied Science and Engineering Technology. 2021;**9**(9):1820-1826. DOI: 10.22214/ijraset.2021.38235

[11] Harish V, Tewari D, Gaur M, Yadav AB, Swaroop S, Bechelany M, et al. Review on nanoparticles and nanostructured materials: Bioimaging, biosensing, drug delivery, tissue engineering, antimicrobial, and agro-food applications. Nanomaterials. 2022;**12**(3):1-43. DOI: 10.3390/nano12030457

[12] Nagamune T. Biomolecular engineering for nanobio/bionanotechnology. Nano Convergence. 2017;**4**(9):1-56. DOI: 10.1186/s40580-017-0103-4

[13] Dosanjh M. From particle physics to medical applications. In: From Particle Physics to Medical Applications. IOP Publishing; 2017;**2399-2891**:1-21. DOI: 10.1088/97807503-1444-2ch1

[14] Aseri A, Garg SK, Nayak A, Trivedi SK, Ahsan J. Magnetic nanoparticles: Magnetic nano-technology using biomedical applications and future prospects. International

Journal of Pharmaceutical Sciences Review and Research. 2015;**31**(2):119-131

[15] Bailey DL, Humm JL, Todd-Pokropek A, Van Aswegen A. Nuclear Medicine Physics A Handbook for Teachers and Students

[16] Podgoršak EB. Graduate Texts in Physics Radiation Physics for Medical Physicists. 3rd ed

[17] Mengesha WG. Cutting-edge physics driven advancements in medical industry. American Journal of Modern Physics. 2024;**13**(2):27-33. DOI: 10.11648/j.ajmp.241302.13

[18] Ghobashy MM. Ionizing radiation-induced polymerization. Ionizing Radiation Effects and Applications. 2018;**1**:1-10. DOI: 10.5772/intechopen.73234

[19] Nikalje AP. Nanotechnology and its applications in medicine. Medicinal Chemistry. 2015;**5**(2):81-89. DOI: 10.4172/2161-0444.1000247

[20] Sisay Mekonen C. Observation of the development of nuclear science and technology as socio-economic and health problems of countries. Radiation Science and Technology. 2023;**9**(4):41-48. DOI: 10.11648/j.rst.20230904.11

[21] Bardhan N. Nanomaterials in diagnostics, imaging and delivery: Applications from COVID-19 to cancer. MRS Communications. 2022;**12**(6):1119-1139. DOI: 10.1557/s43579-022-00257-7

[22] Pinto-coelho L. How artificial intelligence is shaping medical imaging technology: A survey of innovations and applications. Bioengineering (Basel: MDPI). 2023;**10**:1435. DOI: 10.3390/bioengineering10121435

[23] Hussain S, Mubeen I, Ullah N, Shah SS, Khan BA, Zahoor M, et al. Modern diagnostic imaging technique applications and risk factors in the medical field: A review. Hindawi, BioMed research international. 2022;**2022**:1-19. DOI: 10.1155/2022/5164970

[24] Bortfeld T. IMRT: A review and preview. Physics in Medicine and Biology. 2006;**51**(13):363-379. DOI: 10.1088/0031-9155/51/13/R21

[25] Pesapane F, Codari M, Sardanelli F. Artificial intelligence in medical imaging: threat or opportunity? Radiologists again at the forefront of innovation in medicine. European Radiology Experimental. 2018;**2**:35. DOI: 10.1186/s41747-018-0061-6

[26] Tang X. The role of artificial intelligence in medical imaging research. BJR|Open. 2020;**2**(1):20190031. DOI: 10.1259/bjro.20190031

[27] Buaka ESD, Moid MZI. AI and medical imaging technology: Evolution, impacts, and economic insights. The Journal of Technology Transfer. 2024;**49**(1):1-20. DOI: 10.1007/s10961-024-10100-x

[28] Ou X, Chen X, Xu X, Xie L, Chen X, Hong Z, et al. Recent development in X-ray imaging technology: Future and challenges. Research. 2021;**2021**:1-21. DOI: 10.34133/2021/9892152

[29] Dams ETM, Corstens FHM. Lessons for medicine and nuclear medicine research. European Journal of Nuclear Medicine. 1999;**26**(4):311-313. DOI: 10.1007/s002590050391

[30] Jennings WA. Quantities and units for use in radiation protection. Journal of the Society for Radiological Protection. 1982;**2**(2):10-13. DOI: 10.1088/0260-2814/2/2/401

[31] Huang X, El-Sayed MA. Gold nanoparticles: Optical properties and implementations in cancer diagnosis and photothermal therapy. Journal of Advanced Research. 2010;**1**(1):13-28. DOI: 10.1016/j.jare.2010.02.002

[32] Dippong T. Innovative nanomaterial properties and applications in chemistry, physics, medicine, or environment. Nanomaterials (Basel: MDPI). 2024;**14**:145. DOI: 10.3390/nano14020145

[33] Mondini M, Levy A, Meziani L, Milliat F, Deutsch E. Radiotherapy–immunotherapy combinations – Perspectives and challenges. Molecular Oncology. 2020;**14**(7):1529-1537. DOI: 10.1002/1878-0261.12658

[34] Tronchin S, Forster JC, Hickson K, Bezak E. Dosimetry in targeted alpha therapy. A systematic review: Current findings and what is needed. Physics in Medicine and Biology. 2022;**67**(9):09TR01. DOI: 10.1088/1361-6560/ac5fe0

[35] Shannoun F, Blettner M, Schmidberger H, Zeeb H. Strahlenschutz in der diagnostischen radiologie. Deutsches Ärzteblatt. 2008;**105**(3):41-46. DOI: 10.3238/arztebl.2008.0041

[36] Doi K. Computer-aided diagnosis in medical imaging: Historical review, current status and future potential. Computerized Medical Imaging and Graphics. 2007;**31**(4-5):198-211. DOI: 10.1016/j.compmedimag.2007.02.002

[37] Kim J, Hong J, Park H. Foprimer. 2018;**2**(2):37-52. DOI: 10.23838/pfm.2018.00030

[38] Baskar R, Lee KA, Yeo R, Yeoh KW. Cancer and radiation therapy: Current advances and future directions. International Journal of Medical Sciences. 2012;**9**(3):193-199. DOI: 10.7150/ijms.3635

[39] Chen HHW, Kuo MT. Improving radiotherapy in cancer treatment: Promises and challenges. Oncotarget. 2017;**8**(37):62742-62758. DOI: 10.18632/oncotarget.18409

[40] Connell PP, Hellman S. Advances in radiotherapy and implications for the next century: A historical perspective. Cancer Research. 2009;**69**(2):383-392. DOI: 10.1158/0008-5472.CAN-07-6871

[41] Miller CG, Krasnow J, Schwartz LH. Medical Imaging in Clinical Trials. 2014:1-622. ISBN: 9781848827103. DOI: 10. 1007/978-1-84882-710-3

[42] Bárdyová Z, Horváthová M, Pinčáková K, Budošová D. The importance of public health in radiology and radiation protection. Journal of Public Health Research. 2021;**10**(3):1-6. DOI: 10.4081/jphr.2021.2141

[43] Huang P, Tang Q, Li M, Yang Q, Zhang Y, Lei L, et al. Manganese-derived biomaterials for tumor diagnosis and therapy. Journal of Nanobiotechnology. 2024;**22**(335):1-33. DOI: 10.1186/s12951-024-02629-8

[44] Burnet NG, Thomas SJ, Burton KE, Jefferies SJ. Defining the tumour and target volumes for radiotherapy. Cancer Imaging. 2004;**4**(2):153-161. DOI: 10.1102/1470-7330.2004.0054

[45] Roy I. Materials advances therapeutic applications of magnetic nanoparticles: Recent advances. Materials Advances (Cambridge: Royal Society of Chemistry). 2022;**3**:7425-7444. DOI: 10.1039/d2ma00444e

[46] Palma D, Visser O, Lagerwaard FJ, Belderbos J, Slotman BJ, Senan S. Impact of introducing stereotactic

lung radiotherapy for elderly patients with stage I non-small-cell lung cancer: A population-based time-trend analysis. Journal of Clinical Oncology. 2010;**28**(35):5153-5159. DOI: 10.1200/JCO. 2010.30.0731

[47] Gautam B, Parsai EI, Shvydka D, Feldmeier J, Subramanian M. Dosimetric and thermal properties of a newly developed thermobrachytherapy seed with ferromagnetic core for treatment of solid tumors. Medical Physics. 2012;**39**(4):1-10. DOI: 10.1118/1.3693048

[48] Jørgensen SH, Bøgh N, Hansen ESS, Væggemose M, Wiggers H, Laustsen C. Hyperpolarized MRI-an update and future perspectives. Seminars in Nuclear Medicine. 2022;**52**(3):374-381. DOI: 10.1053/j.semnuclmed.2021.09.001

[49] Kraeber-Bodéré F, Barbet J. Challenges in nuclear medicine: Innovative theranostic tools for personalized medicine. Frontiers in Medicine. 2014;**1**(no.Jul):20-22. DOI: 10.3389/fmed.2014.00016

[50] Corradini S, Krug D, Meattini I, Fastner G, Matuschek C, Cutuli B. Challenges in radiotherapy. Breast Care. 2019;**14**(3):152-158. DOI: 10.1159/000500847

[51] Salata OV. Journal of Nanobiotechnology. Journal of Nanobiotechnology. 2004;**6**(3):1-6. DOI: 10.1186/1477-3155-2-12

[52] Fearon K, Arends J, Baracos V. Understanding the mechanisms and treatment options in cancer cachexia. Nature Reviews. Clinical Oncology. 2013;**10**(2):90-99. DOI: 10.1038/nrclinonc.2012.209

[53] Sgouros G, Bodei L, McDevitt MR, Nedrow JR. Radiopharmaceutical therapy in cancer: Clinical advances and challenges. Nature Reviews. Drug Discovery. 2020;**19**(9):589-608. DOI: 10.1038/s41573-020-0073-9

[54] Vaz SC, Oliveira F, Herrmann K, Veit-Haibach P. Nuclear medicine and molecular imaging advances in the 21st century. The British Journal of Radiology. 2020;**93**(1110):1-12. DOI: 10.1259/bjr.20200095

[55] Ribeiro A et al. Ionising radiation exposure from medical imaging – A review of Patient's (un) awareness. Radiography. 2020;**26**(2):e25-e30. DOI: 10.1016/j.radi.2019.10.002

[56] Haleem A, Javaid M, Singh RP, Rab S, Suman R. Applications of nanotechnology in medical field: A brief review. Global Health Journal. 2023;**7**(2):70-77. DOI: 10.1016/j.glohj.2023.02.008

[57] Kaushik D, Jangra P, Verma R, Purohit D, Pandey P, Sharma S, et al. Radiopharmaceuticals: An insight into the latest advances in medical uses and regulatory perspectives. Journal of Biosciences. 2021;**46**(1):1-25. DOI: 10.1007/s12038-021-00147-5

[58] Dhoundiyal S et al. Radiopharmaceuticals: Navigating the frontier of precision medicine and therapeutic innovation. European Journal of Medical Research. 2024;**29**(1):1-19. DOI: 10.1186/s40001-023-01627-0

[59] Hansen SB, Bender D. Advancement in production of radiotracers. Seminars in Nuclear Medicine. 2022;**52**(3):266-275. DOI: 10.1053/j.semnucl med.2021.10.003

[60] International Atomic Energy Agency. Production and Quality Control of Actinium-225 Radiopharmaceuticals. Vienna: IAEA; 2024. pp. 95h3-2ji2. DOI: 10.61092/iaea

[61] Guidoccio F, Mazzarri S, Orsini F, Erba PA, Mariani G. Novel radiopharmaceuticals for therapy. Radiopharmaceuticals: Advances in Therapy (Cham: Springer). 2017. DOI: 10.1007/978-3-319-26236-9_36

[62] Evangel Chinyere Anyanwu, Osasona F, Akomolafe OO, Ogugua JO, Olorunsogo T, Daraojimba ER. Biomedical engineering advances: A review of innovations in healthcare and patient outcomes. International Journal of Science and Research Archive. 2024;**11**(1):870-882. DOI: 10.30574/ijsra.2024.11.1.0139

[63] Damodharan J. Nanomaterials in medicine - an overview. Materials Today Proceedings. 2020;**37**(Part 2):383-385. DOI: 10.1016/j.matpr.2020.05.380

[64] Cheng Z, Li M, Dey R, Chen Y. Nanomaterials for cancer therapy: Current progress and perspectives. Journal of Hematology & Oncology. 2021;**14**(1):1-27. DOI: 10.1186/s13045-021-01096-0

[65] Ng KH, Wong JHD, Yeong CH, Zin HM, Jamal N. Medical physics contributes to the advancement in medicine. ASM Science Journal. 2021;**14**:1-7. DOI: 10.32802/asmscj.2020.505

[66] Dai W et al. Special issue: Novel methods for Labeling Nanomaterials for biological and environmental tracing radiolabeling of Nanomaterials: Advantages and challenges. Frontiers in Toxicology. 2021;**3**(December):1-6. DOI: 10.3389/ftox.2021.753316

Chapter 5

X-Ray Techniques in Analytical Chemistry

Jeremiah Odhek Masime, Peter Munyao Ndangili and Joseph Owuor Lalah

Abstract

X-ray imaging is an important tool for non-destructive analysis, providing insights into internal structures without damaging the sample. Traditional X-ray techniques often require high doses of radiation and are not flexible enough, requiring advanced methods that improve resolution while minimizing exposure. This chapter examines state-of-the-art X-ray imaging techniques, including X-ray fluorescence (XRF) for elemental analysis, X-ray diffraction (XRD) and powder X-ray diffraction (PXRD) for crystal structure determination, and X-ray absorption spectroscopy (XAS), single crystal X-ray diffraction (SCXRD), and X-ray emission spectroscopy (XES), and analyzes their applications, limitations and development potential to advance the field and highlight their transformative impact on scientific and industrial fields. Innovative approaches such as flexible detectors and sophisticated data analysis are well on their way to revolutionizing X-ray imaging. Since its introduction in the 1890s, X-ray imaging has been essential in medical diagnostics and industrial inspections. Continuous optimization and innovation are crucial for further progress. Literature reviews and case studies demonstrate the benefits of techniques such as SCXRD and XES for precise structural analysis, while small-angle X-ray scattering (SAXS) provides valuable insights into molecular and material properties. The chapter highlights the critical role of X-ray imaging in various fields, including materials science, chemistry, biology, forensics and quality control. Closing research gaps and promoting continuous innovation improves the understanding and application of X-ray technologies. The future of X-ray imaging promises enormous breakthroughs in numerous disciplines. This chapter highlights the importance of advancing X-ray imaging techniques to address new challenges.

Keywords: X-ray imaging, XRF, XRD, XAS, XES, SAXS

1. Introduction to X-ray techniques

X-rays were discovered accidentally by Wilhelm Conrad Röntgen in 1895 and revolutionized medical imaging [1]. This breakthrough enabled precise diagnosis of bone fractures, foreign bodies and dental problems. Roentgen's discovery earned him the Nobel Prize in Physics in 1901. Since then, X-rays have become indispensable in various fields such as medicine, industry and security [2]. X-rays are electromagnetic radiation produced when inner orbital electrons are ejected from atoms and other electrons subsequently transition to lower energy levels [3]. Their wavelengths are between 0.01 and 10 nanometers [4]. When X-rays interact with matter, they can be absorbed, scattered or fluorescence induced. Coherent scattering can produce diffraction patterns that are used to analyze the structure of complex materials [5]. X-ray techniques such as radiography, x-ray diffraction and x-ray fluorescence spectrometry use fundamental principles to analyze the structure, morphology, chemistry and elemental composition of a material. These methods include processes such as fluorescence, diffraction, spectroscopy and absorption fine structure and provide detailed insights into materials [6]. By leveraging emission, absorption, scattering, fluorescence and diffraction mechanisms, these techniques provide essential information for various applications in fields such as materials science, biology and forensics. X-ray techniques are versatile and crucial to advancing our understanding of complex materials and their properties [7].

2. Historical development

The discovery of X-rays by Wilhelm Conrad Röntgen in 1895 revolutionized medical imaging. Advances in the early twentieth century greatly improved imaging capabilities [8]. In 1912, Max von Laue's discovery of X-ray diffraction further transformed scientific research by enabling detailed studies of crystal structures and molecular arrangements [9]. Subsequent developments in X-ray spectroscopy, computed tomography (CT) scanning and digital imaging have greatly improved our understanding of materials and biological systems. These techniques are essential for scientific research and enable precise insights into internal structures [10]. Ongoing innovations in instrumentation, data analysis and imaging modalities continue to drive progress in diverse fields, from healthcare to materials engineering [11].

3. Nature and properties

X-rays and ionizing radiation displace tightly bound electrons from atoms and thus produce ions. These electromagnetic waves have wavelengths ranging from about 0.01–10 nanometers and frequencies between 10^{16} and 10^{19} Hz [12]. X-rays interact with matter through processes such as photoelectric absorption, Compton scattering and Thomson/Rayleigh scattering. This results in a loss of energy as it passes through materials [13]. The ability of X-rays to penetrate materials depends on their energy level; lower energy X-rays (about 17 keV) are used in mammography to visualize tissues with lower electron density, while higher energy X-rays (about 80 keV) are used in CT scans of the brain to penetrate denser tissues. This variation in energy levels is essential for effective medical imaging [12]. Despite their ionizing nature, which requires careful handling, X-rays have driven technological advances in healthcare and materials science and played a critical role in modern scientific and technological applications [2, 14].

4. Safety consideration

X-rays are essential tools in medical imaging, industrial inspections and scientific research, providing valuable insights into the internal structure of matter [13]. IR exposure can lead to acute or chronic health problems, including cancer and a shortened lifespan. Excessive X-rays can pose immediate health risks. Long-term exposure increases the risk of cancer and genetic defects [15]. Technological advancements, such as digital radiography and computed tomography, have enabled the development of lower-dose X-ray systems that prioritize image accuracy and safety [16]. Implementing X-ray safety measures, including shielding and adherence to protocols, is crucial for minimizing the risks of radiation exposure while effectively utilizing technological innovations [17].

5. Various applications of X-ray techniques

X-ray techniques are indispensable in medical diagnosis, scientific research, industry and agriculture. They provide detailed internal information about the nature of materials [18]. X-rays play a critical role in improving healthcare, advancing scientific discovery, improving product quality and optimizing agricultural practices [19, 20]. In the medical field, X-rays are essential for diagnosing fractures, tumors and infections and facilitating treatment planning and monitoring. Techniques such as computed tomography (CT) and fluoroscopy provide three-dimensional imaging and real-time visualization during procedures [21]. X-ray crystallography is crucial to scientific research and enables the determination of atomic and molecular structures essential for drug development, materials science and molecular biology studies [22]. Industries rely on X-ray techniques for non-destructive testing, quality control and inspection in aerospace, automotive and electronics manufacturing [23]. In agriculture, X-ray imaging helps assess seed quality, detect foreign bodies in food, and study plant root systems to increase crop production [24].

6. Importance of X-ray techniques in analytical chemistry

X-ray techniques are revolutionizing atomic and molecular studies by providing robust tools for elemental analysis, structure determination and quantitative measurements that are critical in materials science and environmental chemistry [25, 26]. X-ray fluorescence spectroscopy (XRF) and X-ray diffraction (XRD) are powerful techniques used for analyzing material composition and structure. They play essential roles in quality control, environmental monitoring and materials research. XRD includes methods such as powder X-ray diffraction (PXRD) and single crystal X-ray diffraction (SCXRD). They are essential for phase identification, crystallinity determination and structural analysis in areas such as pharmaceuticals, geology and nanotechnology [27]. XRD is a technique used to analyze X-ray scattering from crystal lattices to determine crystal symmetry, unit cell parameters and atomic positions. This data is crucial for identifying crystalline phases and characterizing structural properties, which are essential in materials science, pharmaceuticals and catalysis research [28].

X-ray techniques are crucial in analytical chemistry for quantitative analysis. X-ray absorption spectroscopy (XAS) is valuable for understanding the electronic and chemical properties of atoms, determining oxidation states and coordination geometries [29]. XAS techniques like X-ray absorption near-edge structure (XANES) and extended X-ray absorption fine structure (EXAFS) provide detailed spectral data for analyzing concentrations of specific chemical species and their binding properties [30]. This quantitative capability is essential for studying catalysts, environmental contaminants and biological samples where accurate elemental speciation is critical. As analytical chemistry progresses, X-ray techniques will drive innovation and deepen our understanding of the chemical field [31].

7. Fundamentals of X-ray-matter interaction

X-rays interact with matter in different ways, influencing their behavior upon encountering a substance. High-energy X-rays can excite electrons, as shown in **Figure 1**.

X-rays interact with matter through photoelectric absorption, Compton scattering and coherent scattering. During photoelectric absorption, low-energy X-ray photons eject electrons from the inner shell, creating photoelectrons. Compton scattering occurs at higher energies when X-ray photons are scattered off free or loosely bound electrons, resulting in a wavelength shift. Coherent scattering is a low-energy interaction that results in no loss of energy. Pair formation, a rare interaction at very high energies, results in the formation of an electron-positron pair near the nucleus [2].

X-rays interact with matter through attenuation, scattering and absorption processes. Understanding these interactions is crucial for applications in medical imaging, material analysis and radiation therapy [13]. This knowledge also helps improve X-ray techniques and interpret X-ray data in analytical chemistry and other scientific fields.

8. X-ray absorption and emission processes

X-ray absorption and emission processes provide insights into X-ray interactions. X-ray absorption spectroscopy (XAS) studies empty orbitals, while X-ray emission spectroscopy (XES) studies occupied orbitals to reveal local electronic structures. After X-ray absorption, characteristic X-ray fluorescence photons are emitted, revealing the elemental composition [31, 32]. Techniques such as X-ray absorption near-edge structure (XANES) and extended X-ray absorption fine structure (EXAFS)

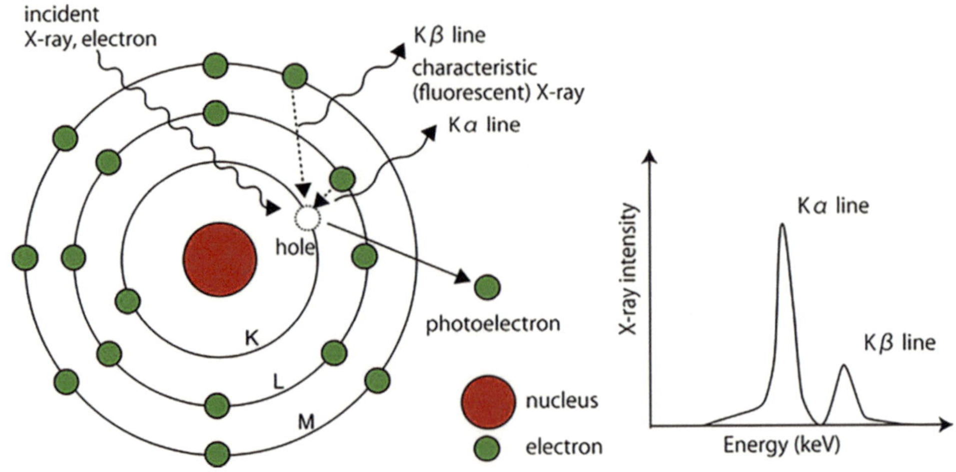

Figure 1.
The process by which distinctive X-rays are produced [5].

analyze energy-dependent absorption to study local electronic and geometric structures. These methods are essential in analytical chemistry, materials science and related fields [29, 33, 34].

9. X-ray scattering phenomena

X-ray scattering is the process by which X-rays can be redirected or deflected when they interact with matter [2]. Compton scattering, discovered in 1923, occurs when X-ray photons are scattered by free or loosely bound electrons in the material [35]. This process is illustrated in **Figure 2**.

The Compton effect is a phenomenon in which X-rays or gamma photons transfer energy and momentum to electrons, causing the scattered photons to have longer wavelengths and lower energies. This interaction is significant in fields such as materials science and medical imaging, where knowledge of radiation-matter interactions is essential. In X-ray crystallography, measuring Compton scattering with polar (θ) and azimuthal (φ) angles is crucial for accurately determining crystal structures [36]. The accuracy of X-ray diffraction (XRD) and X-ray fluorescence (XRF) analyses may be impacted by background noise caused by the wavelength shift from Compton scattering [37].

At low X-ray energies, coherent (Rayleigh) scattering involves the elastic scattering of X-ray photons by bound electrons without energy loss. This process is depicted in **Figure 3**.

Rayleigh scattering geometry was investigated using twisted light and a mesoscopic atomic target of size σ. The quantization axis (z) is aligned with the incident beam direction, and the atomic target is centered on the beam axis. The emission direction of outgoing photons is defined by angle θ_k2, while angle χ represents the polarization vector, enhancing our comprehension of the scattering process [38].

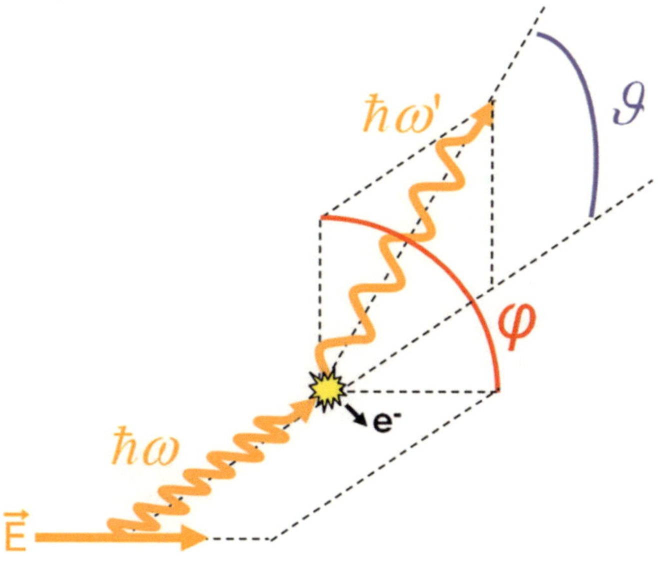

Figure 2.
Compton scattering phenomenon [36].

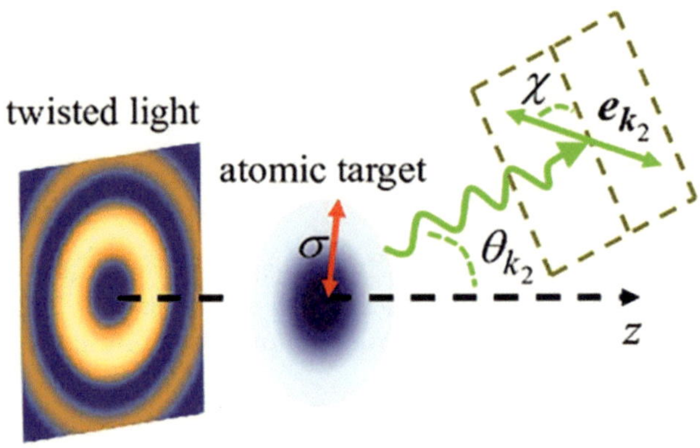

Figure 3.
Geometry of the Rayleigh scattering [38].

10. The phenomenon of X-ray scattering at small angles

The experimental setup for small-angle X-ray scattering (SAXS) is shown in **Figure 4**. A collimated monochromatic light beam is directed onto a scattering cell containing the sample.

The detector measures the scattering intensity at different angles θ from a certain distance. An active beam stop records the intensity of the transmitted beam for normalization while protecting the detector from the primary beam. The scattered wave vector, q, is measured in reciprocal units of length ($Å^{-1}$ or nm^{-1}) and is calculated as the difference between the scattered (kF) and incident (k0) wave vectors. It can be expressed as $q = |kF - k0| = (4\pi/\lambda)\sin(\theta)$ [39]. This provides insights into atomic arrangements, particle size, shape and distribution across various materials, such as liquid dispersions, thin solid objects and powder samples. SAXS operates in the length range of 1–500 nm and is particularly suitable for analyzing complex molecules, polymers and porous materials [40].

Small-angle X-ray scattering (SAXS) is a versatile technique for determining protein structures and understanding molecular building blocks. It quantifies

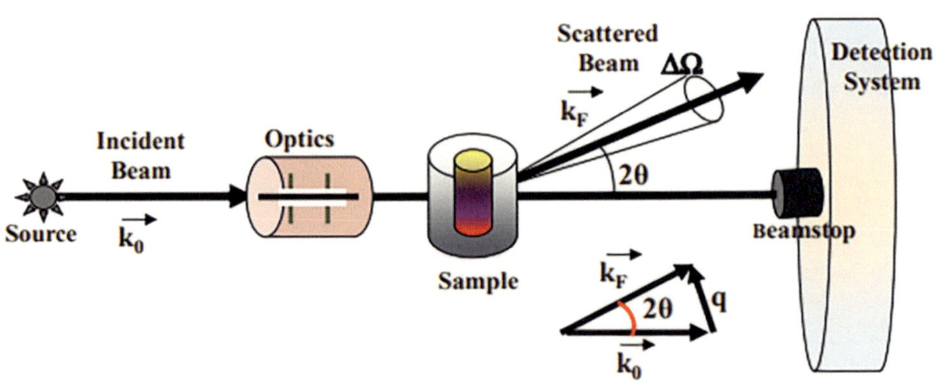

Figure 4.
Schematic of a typical small-angle scattering (SAS) experiment geometry [39].

density fluctuations in nanoscale samples and accurately classifies lipid phases [40, 41]. Recent advances have expanded the application of SAXS in food science to the study of biomacromolecules [39]. Wide-angle X-ray scattering (WAXS) is a valuable technique that complements small-angle X-ray scattering (SAXS) by providing detailed insights into atomic structures, interatomic distances and crystal lattice arrangements. It plays a crucial role in determining crystallographic structures and identifying phases in materials science [42]. By identifying defects that can affect performance, WAXS improves understanding of material properties. This technique is widely used in materials science, polymer analysis and pharmaceutical compound characterization to reveal internal structures and defects. In biological research, WAXS elucidates macromolecular organization and provides insights into molecular structures [42–44].

11. X-ray spectroscopic techniques

11.1 The technique of X-ray fluorescence spectroscopy

Analysis using X-ray fluorescence spectroscopy (XRF) quantifies the amount of X-rays generated when a sample is exposed to high-energy X-ray radiation. This method is similar to optical emission and atomic absorption spectrometry and is used for elemental analysis of solid samples. **Figure 5** shows the results of the XRF spectroscopy.

When X-rays excite electrons in the inner orbitals of a sample, they are ejected, resulting in distinctive X-ray emissions known as K_α and K_β. These emissions occur when an electron from a higher shell (L or M) replaces a K-shell electron that has been ejected. The energies of these X-rays are specific to the elements present in the sample [46]. XRF analysis measures the energy and intensity of X-rays to determine elemental composition. It is based on the unique X-ray emissions of each element, enabling fast and non-destructive analysis in fields such as metallurgy, geology, environmental

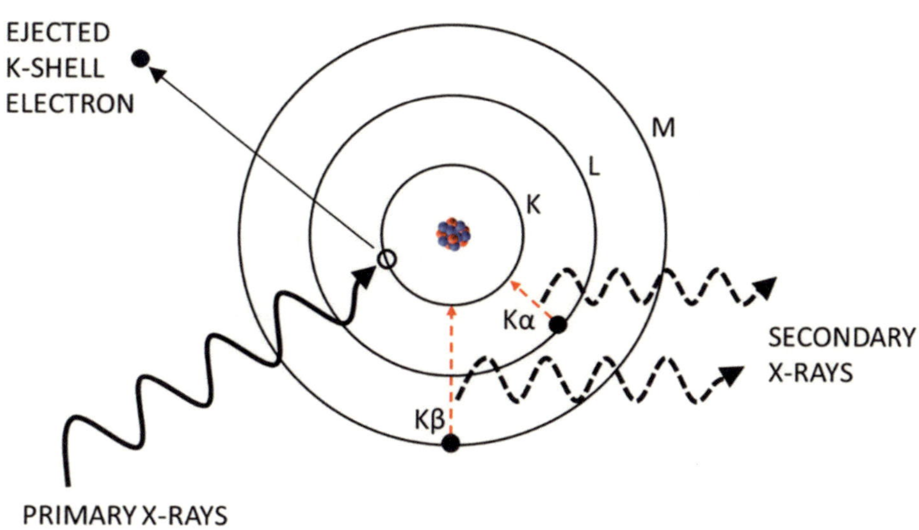

Figure 5.
The physical mechanisms that underlie X-ray fluorescence [45].

sciences and archaeology. The unique emission spectra of each element provide qualitative and quantitative elemental compositions [47].

11.2 The X-ray fluorescence spectroscopy technique

X-ray fluorescence spectroscopy (XRF) uses special equipment that consists of a detector system, a sample chamber and an X-ray source. **Figure 6** shows the essential components of an XRF analyzer in a benchtop setup.

X-ray photoelectron spectroscopy (XPS) uses X-ray photons to remove core electrons from a sample, enabling the analysis of its composition and chemical environment based on the emitted electron energy. By directing high-energy X-rays at the sample, characteristic X-rays are emitted, revealing the elemental composition. XPS provides valuable information about the material being studied [13]. Detectors such as energy or wavelength dispersive detectors capture and analyze these emitted X-rays, enabling rapid and precise elemental analysis. Components such as collimators, filters and software control the X-ray beam, reduce background radiation and help with data collection and interpretation. This instrumentation enables XRF to efficiently quantify elements for applications in materials science, environmental analysis and industrial quality control [49].

The elemental analysis of solid, liquid and powder samples can be done quickly and accurately with X-ray fluorescence spectroscopy (XRF) [48]. XRF is a versatile and non-destructive technique widely used in mining, environmental monitoring and materials science. It is employed for quality control, optimization and research on metals, alloys, ceramics and polymers [27]. Applications of environmental science include assessing contamination levels in soil, sediment and water samples, studying geological formations and monitoring pollutants [50]. XRF also helps identify pigments, metals and materials in cultural artifacts and supports restoration efforts in archaeology and art conservation [51]. By analyzing elemental compositions, it ensures regulatory compliance in the pharmaceutical and food industries. The versatility of XRF extends to forensics, agriculture, geology and semiconductor manufacturing, underscoring its importance for both scientific research and industrial applications.

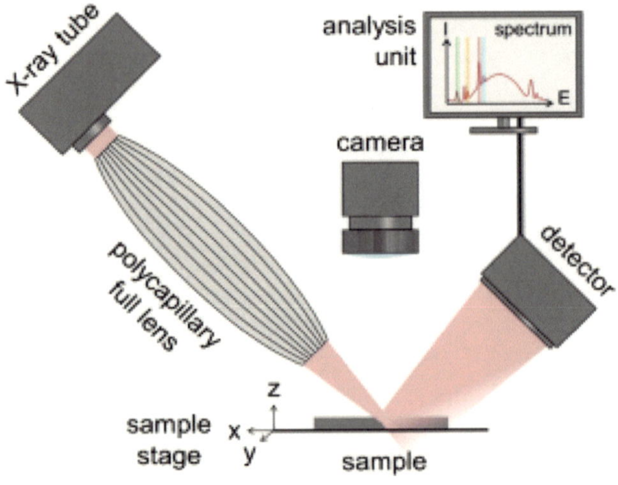

Figure 6.
Schematic representation of a simple benchtop XRF analyzer [48].

11.3 The X-ray photoelectron spectroscopy technique

The elemental composition and chemical states of surfaces are analyzed using X-ray photoelectron spectroscopy (XPS) and electron spectroscopy for chemical analysis (ESCA). These methods provide valuable information about a material's surface properties. **Figure 7** shows the schematic setup for XPS analysis.

X-ray photoelectron spectroscopy (XPS) measures the binding energies of core electrons with an accuracy of ±0.1 eV and high precision. It uses X-rays to eject electrons from the surface of a sample to analyze the elemental composition and chemical state of the material. This method offers insights into the material's elemental composition and chemical states, enhancing our understanding of its properties [51–53]. XPS instruments analyze emitted photoelectrons, allowing for detailed characterization of surface atoms and electronic structures, typically at depths of 1–20 nm [54]. It offers comprehensive surface analysis and chemical information across various disciplines.

In materials science and engineering, XPS is used to characterize elemental composition, chemical bonding, surface chemistry, identify oxidation states, detect contamination and investigate material interfaces [55]. XPS analysis is also employed in catalysis to study surface chemistry, semiconductor quality control, contamination detection and environmental studies for analyzing composition and pollution sources [56, 57]. Additionally, XPS finds application in biomedical research, forensics, archaeology, art conservation and biomaterial analysis, demonstrating its versatility and precision.

11.4 The X-ray absorption spectroscopy technique

X-ray absorption spectroscopy (XAS) provides element-specific information by using X-ray photons to excite core electrons in a sample [32]. The absorption edge has core holes and distinct absorption features, which are due to a significant increase in the absorption coefficient in the core binding energy. XAS creates core holes in the inner shell electrons, creating characteristic absorption edges that help understand

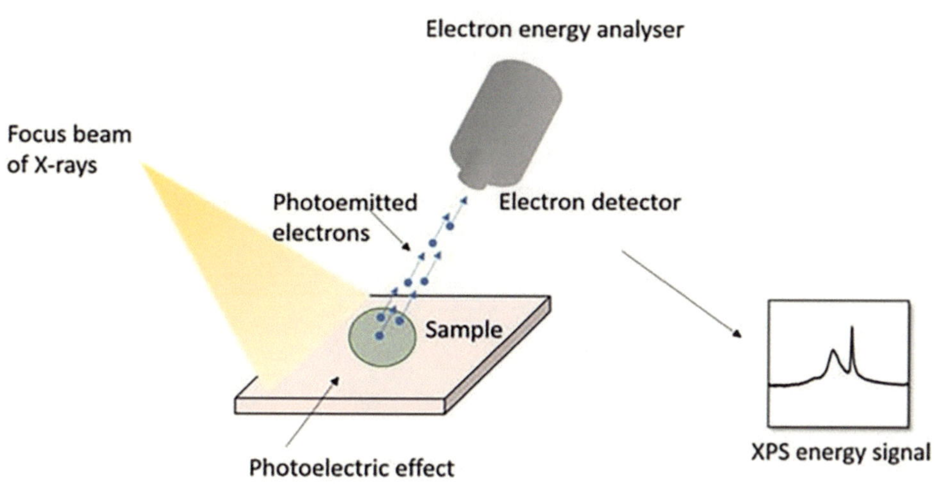

Figure 7.
The working principle of X-ray photoelectron spectroscopy [52].

material composition [58]. By analyzing these edges, researchers can determine the electronic structure, oxidation state, coordination environment and chemical bonding of the elements in the sample [59]. This basic principle enables in-depth studies of atomic structure and electronic properties in various disciplines, including chemistry, physics, materials science and environmental science [60].

11.5 The near-edge and extended X-ray absorption fine structure spectroscopy techniques

X-ray absorption spectroscopy (XAS) includes X-ray absorption near-edge structure spectroscopy (XANES) and extended X-ray absorption fine structure spectroscopy (EXAFS). XAS is a powerful method for studying electronic configurations and molecular structures, providing information about the different valence and spin states of elementary sites. This versatility makes it ideal for analyzing complex systems.

11.6 The X-ray absorption near-edge structure spectroscopy technique

X-ray absorption near-edge structure (XANES) is a method used to analyze the electronic and chemical states of an element by examining the spectral region near its absorption edge. This technique provides important information about how elements behave under varying conditions, particularly in relation to alterations in electronic structure caused by environmental factors. **Figure 8** illustrates the principles and setup for *in situ* XANES analysis.

X-ray absorption near-edge structure analysis (XANES) is used to determine the oxidation state, coordination environment and chemical bonding in a sample and provides valuable insights into electronic structure changes [62]. When X-ray photons interact with a sample, they can be absorbed by electrons in the inner shell, resulting in characteristic absorption edges [63]. XANES analysis focuses on the absorption

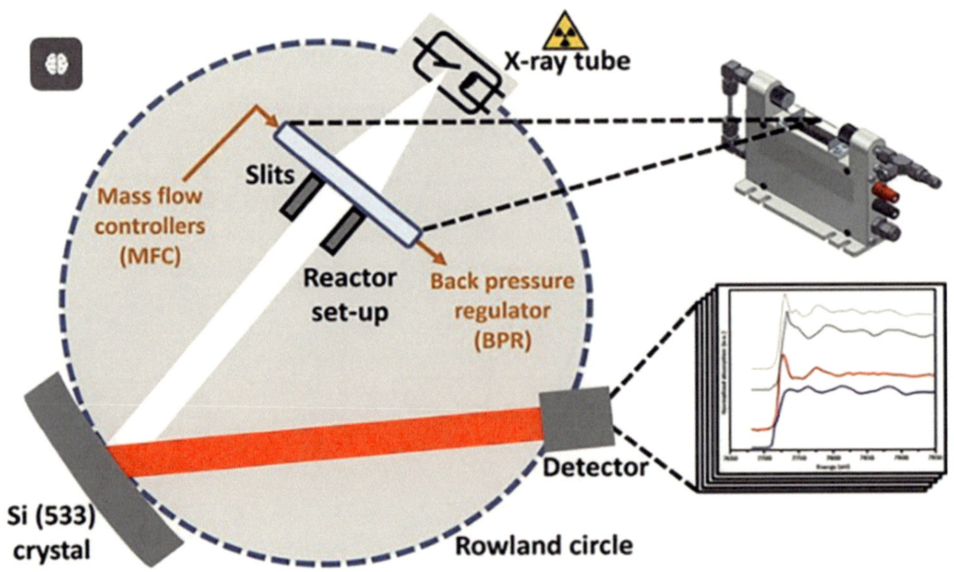

Figure 8.
An illustrated illustration of the In Situ *XANES design, principles and configuration [61].*

spectra near these edges and provides information about the electronic and chemical states [58]. This analysis method helps in studying element speciation and structure-property relationships in various systems, providing detailed information about the local electronic structure. It is widely used in biology, chemistry, materials science and environmental science [64].

11.7 The extended X-ray absorption fine structure spectroscopy technique

X-ray absorption spectroscopy uses the extended X-ray absorption fine structure (EXAFS) technique to analyze oscillatory changes in X-ray absorption beyond the absorption edge to study the local structure around specific elements. EXAFS provides detailed insights into these variations [65]. XAS techniques use X-rays in a specific energy range to measure the intensities of incident and transmitted X-rays, following Beer's law. Absorption edges occur when the incident energy matches the electron binding energy, allowing for the identification and analysis of elements in the sample [64]. XANES spectra derived from EXAFS data offer insights into the coordination environment and average oxidation state of samples [66]. The EXAFS technique uses detectors placed perpendicular to the sample to capture X-ray absorption signals in total fluorescence yield mode. This setup enables the analysis of distances, coordination numbers and bond lengths between atoms, offering valuable insights into the local structural environment of the elements under study [65]. This makes EXAFS an indispensable method for exploring local atomic structures in various materials.

11.8 The X-ray absorption spectroscopy technique

X-ray absorption spectroscopy (XAS) is a flexible technique used in industries and scientific research to examine the chemical environments, electronic characteristics and atomic structures of materials [67]. Its applications range from catalyst optimization in petroleum refining to environmental remediation. XAS characterizes materials such as metals, ceramics and nanoparticles and provides insights into their composition, bonding and structure. In environmental research, XAS helps assess the speciation and mobility of elements, thereby supporting the assessment and remediation of environmental pollution [68]. In geology, XAS illuminates mineral properties, geochemical processes and planetary systems [30]. In addition, XAS studies metalloproteins and enzymes and sheds light on their coordination geometry and function in biological processes such as photosynthesis and metal ion homeostasis [66].

11.9 The X-ray diffraction spectroscopy technique

11.9.1 The principle of XRD spectroscopy

X-ray diffraction (XRD) is the process through which X-rays interact with crystalline materials to reveal structural information. It involves interference in the crystal lattice that can be constructive or destructive. The schematic of the X-ray diffraction process is shown in **Figure 9**.

X-ray diffraction (XRD) equipment consists of essential components such as an X-ray source, sample stage, monochromator, detector and data analysis software. These components work together to produce, interact with and detect X-rays for the analysis of crystalline materials [70]. X-ray diffraction (XRD) uses a monochromatic X-ray beam to analyze a sample's crystal lattice. Detectors record diffracted X-ray

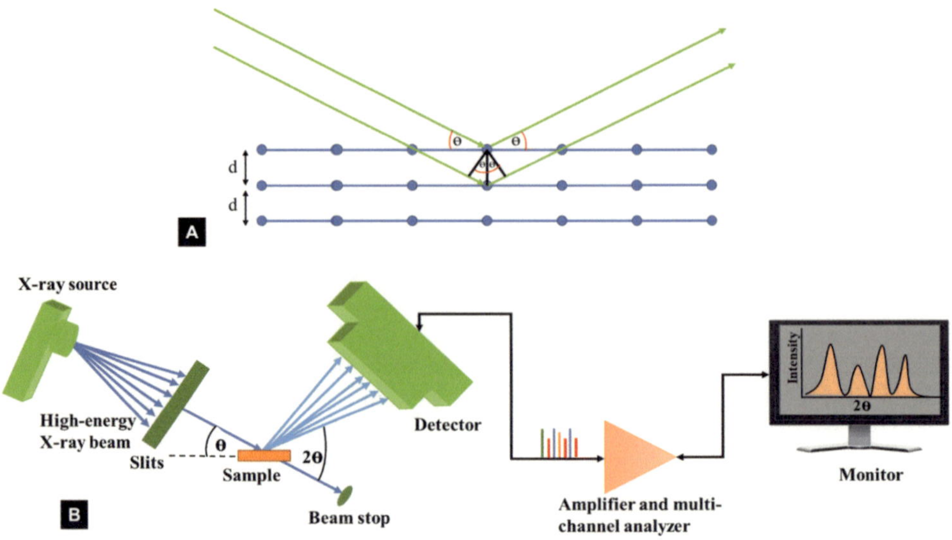

Figure 9.
Schematic representation of Bragg's law (A) and typical XRD instrumentation setup for the powder diffractometer (B) [69].

intensity at different angles, providing insights into the material's properties, composition and structure [71]. Monochromatic X-rays with a wavelength of 1.5406 Å are generated by copper X-ray tubes for analyzing inorganic materials. The data analysis software processes raw diffraction data by subtracting background, identifying peaks, fitting and conducting quantitative analysis to provide structural information on crystalline materials [69].

X-ray diffraction (XRD) is a crucial analytical method in chemistry for determining the structure, composition and properties of materials [28]. Phase identification is critical for forensic analysis and quality control. XRD is a valuable tool for the quantitative analysis of crystalline phases in mixtures to determine the purity of industrial and pharmaceutical products [40]. XRD provides detailed structural information, including lattice parameters and atomic arrangements, which are critical for material characterization and understanding material behavior [69]. Additionally, it helps evaluate material integrity and reliability in various applications by providing insights into texture, anisotropic properties and stress-strain analysis [72].

12. Techniques for X-ray imaging

12.1 X-ray radiography technique

X-ray radiography uses X-rays to create 2D images of internal structures by passing them through objects to a detector [19]. Techniques like lead beam stoppers and posterior beam stop (PBS) matrices reduce scatter and improve image sharpness, while high-resolution charge-coupled device (CCD) arrays support detailed analysis [73]. It is widely used in medical procedures such as fluoroscopy, dental radiography and mammography, as well as for diagnosing fractures. In industries, X-ray Non-destructive testing (NDT) is essential for defect identification and quality assurance, playing a critical role in healthcare and manufacturing [74]. X-ray radiography is also crucial for diagnosis and treatment in forensic science, archaeological investigations and veterinary medicine [75]. It is essential for diagnostics and research due to its adaptability.

12.2 The X-ray computed tomography technique

X-ray computed tomography (CT) uses ionizing radiation to create detailed cross-sectional images by interacting with atoms in an object [76]. Higher-density materials have higher attenuation rates compared to lower-density materials, which affects image quality depending on the material thickness [77]. As the object rotates, X-rays penetrate and are detected, allowing the reconstruction of 3D images through special software and algorithms [13]. In the medical field, X-ray CT is used to diagnose disease, detect internal injuries, assess vascular abnormalities and assist in surgical planning [21]. In industrial applications, it is used for non-destructive testing and inspection to ensure quality control in the aerospace and automotive industries [10]. X-ray CT also plays a crucial role in materials science as it analyzes microstructures and properties and supports research in various fields such as geology, biology, archaeology, earth sciences, environmental sciences and cultural heritage conservation as it reveals internal features without causing damage to cause to the object [2, 13].

12.3 The X-ray crystallography technique

X-ray crystallography is an important technique for determining the atomic and molecular structures of crystalline materials by shining X-rays onto a crystal and analyzing the resulting diffraction pattern to create a detailed 3D model of the atomic arrangement [78]. This technique is crucial for understanding material structures, characterizing crystal symmetry, lattice parameters and phase transitions [79]. It is also widely used in drug design, interaction studies and biomolecular research to study proteins, nucleic acids and their complexes [80]. X-ray crystallography also plays an important role in environmental and geological sciences by analyzing minerals, rocks and soils, contributing to geological understanding and environmental remediation [27].

13. The emerging technologies

Recent advancements in X-ray techniques for analytical chemistry have significantly improved the sensitivity, resolution and versatility of material analyses. The integration of X-ray techniques with other imaging modalities, such as optical microscopy and spectroscopy, enables comprehensive material characterization by providing complementary information on chemical composition and structure [81]. High-speed X-ray imaging developments allow real-time monitoring of dynamic processes, while advanced data analysis algorithms facilitate the extraction of quantitative information, improving efficiency and accuracy [82]. *In situ* and operando studies under realistic conditions provide insights into dynamic processes and reaction mechanisms in real time. Synchrotron and free-electron laser sources offer high-brilliance X-ray beams for advanced spectroscopy, diffraction and imaging experiments [83]. Nanoscale imaging and spectroscopy techniques such as scanning transmission x-ray microscopy (STXM) provide detailed visualization of nanomaterials and biological structures, enabling insights into their properties and behavior at the molecular level. These techniques have diverse applications in environmental monitoring, biomedical research and pharmaceutical development [84].

14. Challenges and limitations

X-ray techniques have advanced analytical chemistry by improving materials characterization and imaging. However, they face challenges and limitations that can impact their effectiveness and accessibility. See **Table 1** for a summary.

15. Future prospects in analytical chemistry

The future of X-ray techniques in analytical chemistry looks promising, with advancements in instrumentation, methods and applications expected to enhance sensitivity, spatial resolution and the characterization of trace elements and molecular structures [90]. Integration with other imaging modalities will provide comprehensive insights into material composition and dynamics, while improvements in sample environments will enable real-time study of dynamic processes like chemical reactions and biological interactions [81]. Automation and rapid data processing will expedite material screening and characterization and advanced tomography methods will enhance visualization of material dynamics and morphology. Artificial intelligence and machine learning algorithms will facilitate faster discoveries and innovations by improving data analysis and interpretation [91]. X-ray techniques are increasingly being applied in environmental monitoring, biomedical research and pharmaceutical development to tackle challenges such as pollution cleanup and disease diagnosis [92].

Challenge/Limitations	Description	Implications
Preparation of the sample	Sample preparation often requires extensive work and may involve creating thin sections, powders or crystals [81].	This procedure may limit the types of samples that can be analyzed and introduce artifacts that affect accuracy [81].
Analysis of trace elements	Trace element analysis may not be sensitive enough for traditional X-ray techniques [85].	Researchers are exploring advanced methods, such as synchrotron radiation, to enhance sensitivity and spatial resolution [85].
Damage from Extended Radiation Exposure	Long-term X-ray exposure can damage delicate materials by altering their structure and chemical composition [86].	This limits the examination of sensitive materials and requires careful management of irradiation time [86].
Accessibility and Cost	Synchrotrons and other advanced X-ray equipment are expensive and require specialized expertise to operate [87].	Access to research and institutions is often limited by high costs and specialized knowledge [87].
The complexity of data analysis	Expertise, significant computing resources and advanced software are essential for data analysis [88].	Researchers, especially those in smaller labs or with limited resources, may encounter challenges [88].
Environmental and safety issues	Strict safety procedures and environmental considerations are essential when using X-ray techniques due to the risks of radiation exposure [89].	Grammatical errors and typos in laboratory operations can lead to challenges and require additional staff supervision and training [89].

Table 1.
Summary of the challenges and limitations of X-ray techniques.

16. Conclusion

X-ray techniques play a crucial role in analytical chemistry for material characterization, structural analysis, environmental monitoring and biological imaging, offering valuable insights into material composition and dynamics. Despite facing challenges such as sample preparation, sensitivity, spatial resolution and data complexity, advancements in instrumentation and machine learning are expected to improve their capabilities. Collaborations across disciplines will drive innovation, tackle scientific and social challenges, and advance science for the betterment of humanity.

Acknowledgements

I am grateful to my colleagues at the School of Chemistry and Material Science, as well as the library staff and laboratory technicians at the Technical University of Kenya for their support in providing essential resources for this study. Their assistance was invaluable throughout the research process.

Potential conflicts of interest

There are no conflicts of interest for the author that might have affected the findings of the study.

Author details

Jeremiah Odhek Masime*, Peter Munyao Ndangili and Joseph Owuor Lalah
School of Chemistry and Material Science, The Technical University of Kenya, Nairobi, Kenya

*Address all correspondence to: masimejo@yahoo.com

References

[1] Ige TA, Hospital N, Hasford F, Tabakov S, Trauernicht CJ. Medical Physics Development In Africa - Status, Education, Challenges, Future; 2020. pp. 231-319

[2] Berger AM, Yang Q, Maier A. X-ray imaging. In: Maier A, Steidl S, Christlein V, Hornegger J, editors. Medical Imaging Systems: An Introductory Guide. Springer, Cham: Switzerland; 2018. pp. 119-145

[3] Feng L, Yijin L, Xiqian Y, Lei C, Andrej S, Oleg GS, et al. Synchrotron X-ray analytical techniques for studying materials electrochemistry in rechargeable batteries. Chemical Reviews. 2017;**117**(21):13123-13186. DOI: 10.1021/acs.chemrev.7b00007

[4] Susi T. Other spectroscopic methods for graphene characterization: X-ray and electron spectroscopies. Graphene. 2021;**January**:413-436. DOI: 10.1016/b978-0-08-102848-3.00011-6

[5] Uo M, Wada T, Sugiyama T. Applications of X-ray fluorescence analysis (XRF) to dental and medical specimens. Japanese Dental Science Review. 2015;**51**(1):2-9. DOI: 10.1016/j.jdsr.2014.07.001

[6] Sanchez-Cano C, Alvarez-Puebla RA, Abendroth JM, Beck T, Blick R, Cao Y, et al. X-ray-based techniques to study the nano-bio interface. ACS Nano. 2021;**15**(3):3754-3807. DOI: 10.1021/acsnano.0c09563

[7] Fetisov GV. X-ray diffraction methods for structural diagnostics of materials: Progress and achievements. Uspekhi Fizicheskikh Nauk. 2020;**190**(01):2-36. DOI: 10.3367/ufnr.2018.10.038435

[8] Du Boulay G. Imaging in medicine through the 20th century. Journal of the Royal College of Physicians of London. 2000;**34**(4):357-362

[9] Mingos DMP. Early history of X-ray crystallography. Structure and Bonding. 2020;**185**:1-42. DOI: 10.1007/430_2020_73

[10] Omori NE, Bobitan AD, Vamvakeros A, Beale AM, Jacques SDM. Recent developments in X-ray diffraction/scattering computed tomography for materials science. Philosophical Transactions of the Royal Society A - Mathematical Physical and Engineering Sciences. 2023;**381**(2259):1-29. DOI: 10.1098/rsta.2022.0350

[11] Javaid M, Haleem A, Singh RP. Health informatics to enhance the healthcare industry's culture: An extensive analysis of its features, contributions, applications and limitations. Health Informatics. 2024;**1**(2):123-148. DOI: 10.1016/j.infoh.2024.05.001

[12] Giannini C, Ladisa M, Altamura D, Siliqi D, Sibillano T, De Caro L. X-Ray Diffraction: A Powerful Technique for the Multiple-Length-Scale Structural Analysis of Nanomaterials2016. pp. 1-22. DOI: 10.3390/cryst6080087

[13] Ou X, Chen X, Xu X, Xie L, Chen X, Hong Z, et al. Recent development in X-ray imaging technology: Future and challenges. Research. 2021;**2021**:1-18. DOI: 10.34133/2021/9892152

[14] Anwar M, Abdul MAS, Khan UM, Hassan M, Khoja AH, Muchtar A. A review of X-ray photoelectron spectroscopy technique to analyze the stability and degradation mechanism of solid oxide fuel cell cathode materials. Materials (Basel). 2022;**15**(7):1-23. DOI: 10.3390/ma15072540

[15] Talapko J, Talapko D, Katalini´ D, Kotris I, Eri´ I, Beli´ D, et al. Health effects of ionizing radiation on the human body. Medicina (B. Aires). 2024;**60**(4):653. DOI: 10.3390/MEDICINA60040653

[16] Lakhwani OP, Dalal V, Jindal M, Nagala A. Radiation protection and standardization. Journal of Clinical Orthopaedics and Trauma. 2019;**10**(4):738-743. DOI: 10.1016/j.jcot.2018.08.010

[17] Ahmed TAN, Taha S. Radiation exposure, the forgotten enemy: Toward implementation of national safety program. The Egyptian Heart Journal. 2017;**69**(1):55-62. DOI: 10.1016/j.ehj.2016.10.001

[18] Mathanker SK, Weckler PR, Wang N. Terahertz (THZ) applications in food and agriculture: A review. Transactions of the ASABE. 2013;**56**(3):1213-1226. DOI: 10.13031/trans.56.9390

[19] Kotwaliwale N, Singh K, Kalne A, Jha SN, Seth N, Kar A. X-ray imaging methods for internal quality evaluation of agricultural produce. Journal of Food Science and Technology. 2014;**51**(1):1-15. DOI: 10.1007/s13197-011-0485-y

[20] Min H, Lim YK, Park JW. Integrating X-ray technologies with intelligent transportation systems for enhancing the international maritime security. International Journal of Logistics Systems and Management. 2015;**22**(1):1-14. DOI: 10.1504/IJLSM.2015.070888

[21] Hussain S, Mubeen I, Ullah N, Shah SSD, Khan BA, Zahoor M, et al. Modern diagnostic imaging technique applications and risk factors in the medical field: A review. BioMed Research International. 2022;**2022**:1-19. DOI: 10.1155/2022/5164970

[22] Rodríguez I, Gautam R, Tinoco AD. Using x-ray diffraction techniques for biomimetic drug development, formulation, and polymorphic characterization. Biomimetics. 2021;**6**(1):1-23. DOI: 10.3390/biomimetics6010001

[23] Zhang X, Cheng L, Liu Y, Tao B, Wang J, Liao R. A review of non-destructive methods for the detection tiny defects within organic insulating materials. Frontiers in Materials. 2022;**9**(August):1-15. DOI: 10.3389/fmats.2022.995516

[24] Du Z, Hu Y, Ali Buttar N, Mahmood A. X-ray computed tomography for quality inspection of agricultural products: A review. Food Science and Nutrition. 2019;**7**(10):3146-3160. DOI: 10.1002/fsn3.1179

[25] Prizes N, X-rays B, X-ray C. Introduction: X-rays in chemistry. Chemical Reviews. 2001;**101**(6):1-2

[26] Aitipamula S, Vangala VR. X-ray crystallography and its role in understanding the physicochemical properties of pharmaceutical cocrystals. Journal of the Indian Institute of Science. 2017;**97**(2):227-243. DOI: 10.1007/s41745-017-0026-4

[27] Oyedotun TDT. X-ray fluorescence (XRF) in the investigation of the composition of earth materials: A review and an overview. Geology, Ecology, and Landscapes. 2018;**2**(2):148-154. DOI: 10.1080/24749508.2018.1452459

[28] Abid HR, Azhar MR, Iglauer S, Rada ZH, Al-Yaseri A, Keshavarz A. Physicochemical characterization of metal organic framework materials: A mini review. Heliyon. 2024;**10**(1):e23840. DOI: 10.1016/j.heliyon.2023.e23840

[29] Harper AF, Emge SP, Magusin PCMM, Grey CP, Morris AJ. Modelling amorphous materials via a joint solid-state NMR and X-ray absorption spectroscopy and DFT approach: Application to alumina. Chemical Science. 2022;**14**(5):1155-1167. DOI: 10.1039/d2sc04035b

[30] Mastelaro VR, Zanotto ED. X-ray absorption fine structure (XAFS) studies of oxide glasses-a 45-year overview. Materials (Basel). 2018;**11**(2):1-40. DOI: 10.3390/ma11020204

[31] Llaver M, Fiorentini EF, Oviedo MN, Quintas PY, Wuilloud RG. Elemental speciation analysis in environmental studies: Latest trends and ecological impact. International Journal of Environmental Research and Public Health. 2021;**18**(22):1-27. DOI: 10.3390/ijerph182212135

[32] Zimmermann P, Peredkov S, Abdala PM, DeBeer S, Tromp M, Müller C, et al. Modern X-ray spectroscopy: XAS and XES in the laboratory. Coordination Chemistry Reviews. 2020;**423**:213466. DOI: 10.1016/j.ccr.2020.213466

[33] Jönsson O. Ultrafast Structural and Electron Dynamics in Soft Matter Exposed to Intense X-ray Pulses. Uppsala: Acta Universitatis Upsaliensis; 2017. pp. 1-78, diva2:1152670

[34] Loy ACM, Ng WL, Bhattacharya S. Advanced characterization techniques for the development of subatomic scale catalysts: One step closer to industrial scale fabrication. Materials Today Catalysis. 2024;**4**(November 2023):100033. DOI: 10.1016/j.mtcata.2023.100033

[35] Compton AH. A quantum theory of the scattering of X-rays by light elements. Physics Review. 1923;**21**(5):483-502. DOI: 10.1103/PhysRev.21.483

[36] Weber G, Bräuning H, Surzhykov A, Brandau C, Fritzsche S, Geyer S, et al. Combined linear polarization and angular distribution measurements of x-rays for precise determination of multipole-mixing in characteristic transitions of high-Z systems. Journal of Physics B: Atomic, Molecular and Optical Physics. 2015;**48**(14):144031. DOI: 10.1088/0953-4075/48/14/144031

[37] You H. X-ray scattering and imaging studies of electrode structure and dynamics. Chemical Record. 2019;**19**(7):1220-1232. DOI: 10.1002/tcr.201800083

[38] Peshkov AA, Volotka AV, Surzhykov A, Fritzsche S. Rayleigh scattering of twisted light by hydrogenlike ions. Physical Review A. 2018;**97**(2):1-9. DOI: 10.1103/PhysRevA.97.023802

[39] Lombardo D, Calandra P, Kiselev MA. Structural characterization of biomaterials by means of small angle x-rays and neutron scattering (Saxs and sans), and light scattering experiments. Molecules. 2020;**25**(23):1-38. DOI: 10.3390/molecules25235624

[40] Mourdikoudis S, Pallares RM, Thanh NTK. Characterization techniques for nanoparticles: Comparison and complementarity upon studying nanoparticle properties. Nanoscale. 2018;**10**(27):12871-12934. DOI: 10.1039/c8nr02278j

[41] Das A, Yadav N, Manchala S, Bungla M, Ganguli AK. Mechanistic investigations of growth of anisotropic nanostructures in reverse micelles. ACS Omega. 2021;**6**(2):1007-1029. DOI: 10.1021/acsomega.0c04033

[42] Pabit SA, Katz AM, Tolokh IS, Drozdetski A, Baker N, Onufriev AV, et al. Understanding nucleic acid structural changes by comparing wide-angle

x-ray scattering (WAXS) experiments to molecular dynamics simulations. The Journal of Chemical Physics. 2016;**144**(20). DOI: 10.1063/1.4950814

[43] Ochbaum G, Bitton R. Using Small-Angle X-ray Scattering (SAXS) to Study the Structure of Self-Assembling Biomaterials. Elsevier Ltd.; 2018. pp. 291-304. DOI: 10.1016/B978-0-08-102015-9.00015-0

[44] Girão AV, Caputo G, Ferro MC. Application of scanning electron microscopy–energy dispersive X-ray spectroscopy (SEM-EDS). Comprehensive Analytical Chemistry. 2017;**75**(June):153-168. DOI: 10.1016/bs.coac.2016.10.002

[45] Horf M, Gebbers R, Vogel S, Ostermann M, Piepel M, Olfs H. Biogas digestates by portable energy-dispersive X-ray fluorescence spectrometry. Sensors. 2021;**21**:3892, 1-12. DOI: 10.3390/s21113892

[46] Duarte J, Moreira S, Pombo J. Implementation of a methodology for elemental analysis of sediments based on a handheld X-ray fluorescence analyzer. Jornadas de Engenharia Hidrográfica. 2018;**1755**(June):373-376

[47] Yao M, Wang D, Zhao M. Element analysis based on energy-dispersive X-ray fluorescence. Advances in Materials Science and Engineering. 2015;**1**:2015. DOI: 10.1155/2015/290593

[48] Heimler K, Gottschalk C, Vogt C. Confocal micro X-ray fluorescence analysis for the non-destructive investigation of structured and inhomogeneous samples. Analytical and Bioanalytical Chemistry. 2023;**415**(21):5083-5100. DOI: 10.1007/s00216-023-04829-x

[49] Brouwer P. Theory of XRF. Almelo, Netherlands: PANalytical BV; 2010. pp. 1-62

[50] Li S, Shen J, Bishop TFA, Viscarra Rossel RA. Assessment of the effect of soil sample preparation, water content and excitation time on proximal X-ray fluorescence sensing. Sensors. 2022;**22**(12):1-11. DOI: 10.3390/s22124572

[51] Musílek L, Čechák T, Trojek T. X-ray fluorescence in investigations of cultural relics and archaeological finds. Applied Radiation and Isotopes. 2012;**70**(7):1193-1202. DOI: 10.1016/j.apradiso.2011.10.014

[52] Long J, Nand A, Ray S. Application of spectroscopy in additive manufacturing. Materials (Basel). 2021;**14**(1):1-29. DOI: 10.3390/MA14010203

[53] Greczynski G, Hultman L. X-ray photoelectron spectroscopy: Towards reliable binding energy referencing. Progress in Materials Science. 2020;**107**(July):100591. DOI: 10.1016/j.pmatsci.2019.100591

[54] Desimoni E, Brunetti B. X-ray photoelectronspectroscopiccharacterization of chemically modified electrodes used as chemical sensors and biosensors: A review. Chem. 2015;**3**(2):70-117. DOI: 10.3390/chemosensors3020070

[55] Bagus PS, Freund HJ. X-ray photoelectron spectroscopy as a useful tool to study surfaces and model systems for heterogeneous catalysts: A review and perspective. Surface Science. 2024;**745**(January):122471. DOI: 10.1016/j.susc.2024.122471

[56] Krishna DNG, Philip J. Review on surface-characterization applications of X-ray photoelectron spectroscopy (XPS): Recent developments and challenges. Applied Surface Science Advances. 2022;**12**(December). DOI: 10.1016/j.apsadv.2022.100332

[57] Guascito MR, Cesari D, Chirizzi D, Genga A, Contini D. XPS

surface chemical characterization of atmospheric particles of different sizes. Atmospheric Environment. 2015;**116**(June):146-154. DOI: 10.1016/j.atmosenv.2015.06.028

[58] Yano J, Yachandra VK. X-ray absorption spectroscopy. Photosynthesis Research. 2009;**102**(2):241-254. DOI: 10.1007/s11120-009-9473-8

[59] Hummer AA, Rompel A. X-ray Absorption Spectroscopy: A Tool to Investigate the Local Structure of Metal-Based Anticancer Compounds in Vivo. 1st ed. Vol. 93. Elsevier Inc.; 2013. DOI: 10.1016/B978-0-12-416596-0.00008-7

[60] Evans J. X-Ray Absorption Spectroscopy for the Chemical and Materials Sciences2018. DOI: 10.1002/9781118676165

[61] Moya-Cancino JG, Honkanen A, van der Eerden AMJ, Schaink H, Folkertsma L, Ghiasi M, et al. In-situ X-ray absorption near edge structure spectroscopy of a solid catalyst using a laboratory-based set-up. ChemCatChem. 2019;**11**(3):1039-1044. DOI: 10.1002/cctc.201801822

[62] Schrapers P, Mebs S, Ilina Y, Warner DS, Wörmann C, Schuth N, et al. The binuclear nickel center in the A-cluster of acetyl-CoA synthase (ACS) and two biomimetic dinickel complexes studied by X-ray absorption and emission spectroscopy. Journal of Physics Conference Series. 2016;**712**(1):1-5. DOI: 10.1088/1742-6596/712/1/012029

[63] Henderson GS, de Groot FMF, Moulton BJA. X-ray absorption near-edge structure (XANES) spectroscopy. Spectroscopic Methods in Mineralogy and Material Sciences. 2014;**February**:75-138. DOI: 10.2138/rmg.2014.78.3

[64] Wang J, Hsu C, Wu T, Chan T, Suen N, Lee J, et al. In situ X-ray spectroscopies beyond conventional X-ray absorption spectroscopy on deciphering dynamic configuration of electrocatalysts. Nature Communications. 2023;**14**(1):1-23. DOI: 10.1038/s41467-023-42370-8

[65] Britz A, Abraham B, Biasin E, van Driel TB, Gallo A, Garcia-Esparza T, et al. Resolving structures of transition metal complex reaction intermediates with femtosecond EXAFS. Physical Chemistry Chemical Physics. 2020;**22**(5):2660-2666. DOI: 10.1039/c9cp03483h

[66] Chatterjee R, Weninger C, Loukianov A, Gul S, Fuller FD, Cheah MH, et al. XANES and EXAFS of dilute solutions of transition metals at XFELs. Journal of Synchrotron Radiation. 2019;**26**(October 2018):1716-1724. DOI: 10.1107/S1600577519007550

[67] Harmand M, Cammarata M, Chollet M, Krygier AG, Lemke HT, Zhu D. Single-shot X-ray absorption spectroscopy at X-ray free electron lasers. Scientific Reports. 2023;**13**(1):1-11. DOI: 10.1038/s41598-023-44196-2

[68] Aucour AM, Sarret G, Blommaert H, Wiggenhauser M. Coupling metal stable isotope compositions and X-ray absorption spectroscopy to study metal pathways in soil-plant systems: A mini review. Metallomics. 2023;**15**(4):1-19. DOI: 10.1093/mtomcs/mfad016

[69] Salem KS, Kasera NK, Rahman A, Jameel H, Habibi Y, Eichhorn SJ, et al. Comparison and assessment of methods for cellulose crystallinity determination. Chemical Society Reviews. 2023;**52**(18):6417-6446. DOI: 10.1039/d2cs00569g

[70] Saraugi SS. X-Ray Diffraction (XRD). Vol. June2024. DOI: 10.13140/RG.2.2.27075.82726

[71] Siddiqui A, Rahman Z, Korang-Yeboah M, Khan MA. Development and validation of X-ray diffraction method for quantitative determination of crystallinity in warfarin sodium products. International Journal of Pharmaceutics. 2015;**493**(1-2):1-6. DOI: 10.1016/j.ijpharm.2015.07.051

[72] Meena PL, Kumar R, Sreenivas K. Structural, Elastic and Magnetic Properties of Spinel Co 3 O 4. 2018. Available from: http://nopr.niscair.res.in/bitstream/123456789/45351/1/IJPAP56%2811%29 890-895.pdf [Accessed: December 18, 2018]

[73] Onnia V, Varjonen V, Lehtimäki M, Lehtokangas M, Saarinen J. Scatter compensation in digital radiography. The European Signal Processing Conference. 2000;**2015-March**(March):1-4

[74] Seibert JA. Projection X-ray imaging: Radiography, mammography, fluoroscopy. Health Physics. 2019;**116**(2):148-156. DOI: 10.1097/HP.0000000000001028

[75] Yitbarek D, Dagnaw GG. Application of advanced imaging modalities in veterinary medicine: A review. Veterinary Medicine: Research and Reports. 2022;**13**(May):117-130. DOI: 10.2147/vmrr.s367040

[76] Villarraga-Gómez H, Herazo EL, Smith ST. X-ray computed tomography: From medical imaging to dimensional metrology. Precis Engineering. 2019;**60**(May 2022):544-569. DOI: 10.1016/j.precisioneng.2019.06.007

[77] Neutze JA. Computed Tomography. 6th ed. Vol. 3. Springer; 2020. pp. 1-400. DOI: 10.1007/978-3-030-22173-7_6

[78] Shi D, Nannenga BL, Iadanza MG, Gonen T. Three-dimensional electron crystallography of protein microcrystals. eLife. 2013;**2**:1-17. DOI: 10.7554/elife.01345

[79] Bunaciu AA, Udriştioiu EG, Aboul-Enein HY. X-ray diffraction: Instrumentation and applications. Critical Reviews in Analytical Chemistry. 2015;**45**(4):289-299. DOI: 10.1080/10408347.2014.949616

[80] Scimeca M, Bischetti S, Lamsira HK, Bonfiglio R, Bonanno E. Energy dispersive X-ray (EDX) microanalysis: A powerful tool in biomedical research and diagnosis. European Journal of Histochemistry. 2018;**62**(1):89-99. DOI: 10.4081/ejh.2018.2841

[81] Zheng H, Handing KB, Zimmerman MD, Shabalin IG, Almo SC, Minor W. X-ray crystallography over the past decade for novel drug discovery -where are we heading next? Expert Opinion on Drug Discovery. 2015;**10**(9):975-989. DOI: 10.1517/17460441.2015.1061991

[82] Najjar R. Redefining radiology: A review of artificial intelligence integration in medical imaging. Diagnostics. 2023;**13**(17):1-25. DOI: 10.3390/diagnostics13172760

[83] Allen AJ. Selected advances in small-angle scattering and applications they serve in manufacturing, energy and climate change. Journal of Applied Crystallography. 2023;**56**(Pt 3):787-800. DOI: 10.1107/S1600576723003898

[84] Huxford T. X-ray crystallography. In: Brenner's Encyclopedia of Genetics. 2nd ed. 2013;**7**:366-368. DOI: 10.1016/B978-0-12-374984-0.01657-0

[85] Lu H, Campbell E, Campbell DE, Wang C, Ren H. EPA public access. Atmospheric Environment. 2017;**23**(1994):248-258. DOI: 10.1016/j.hal.2017.06.001.Submit

[86] Davis AM, St-Gallay SA, Kleywegt GJ. Limitations and lessons in the use of X-ray structural information in drug design. Drug Discovery Today. 2008;**13**(19-20):831-841. DOI: 10.1016/j.drudis.2008.06.006

[87] Staufer T, Grüner F. Review of development and recent advances in biomedical X-ray fluorescence imaging. International Journal of Molecular Sciences. 2023;**24**(13)1-10. DOI: 10.3390/ijms241310990

[88] Gaudêncio SP, Bayram E, Bilela LL, Cueto M, Díaz-Marrero AR, Haznedaroglu BZ, et al. Advanced methods for natural products discovery: Bioactivity screening, dereplication, metabolomics profiling, genomic sequencing, databases and informatic tools, and structure elucidation. Marine Drugs. 2023;**21**(5):308, 1-66. DOI: 10.3390/md21050308

[89] Alyami J, Nassef MH. Assessment of diagnostic radiology facilities technical radiation protection requirements in KSA. Applied Sciences. 2022;**12**(14):1-10. DOI: 10.3390/app12147284

[90] Thompson MC, Yeates TO, Rodriguez JA. Advances in methods for atomic resolution macromolecular structure determination. F1000Research. 2020;**9**:1-18. DOI: 10.12688/f1000research.25097.1

[91] Feng S, Liotti E, Grant PS. X-ray imaging of alloy solidification: Crystal formation, growth, instability and defects. Materials (Basel). 2022;**15**(4):1-27. DOI: 10.3390/ma15041319

[92] Zheng H, Hou J, Zimmerman MD, Wlodawer A, Minor W. The future of crystallography in drug discovery. Expert Opinion on Drug Discovery. 2014;**9**(2):125-137. DOI: 10.1517/17460441.2014.872623

Chapter 6

Ultra-High Dose Rate Ionizing Radiation for FLASH Radiotherapy Applications

Livia Sima, Monica Tudor, Stefana Orobeti, Anca Filimon, Gabriela Negroiu, Cristian Munteanu, Gabriela Chiritoiu, Anca Roseanu, Paula Florian, Madalina Icriverzi, Ioana Porosnicu, Daniel Avram, Alexandra Bran, Sandel Simion, Ion Tiseanu and Felix Sima

Abstract

Radiation therapy (RT) is an essential component of curative cancer treatment. Technological advances in RT aim to make treatments more precise, efficient, and personalized. Recent FLASH-induced radiotherapeutic strategies employ ultra-high dose rate regimes to overcome toxicity of healthy tissues, maintaining similar efficacy in altering tumor progression with respect to conventional dose rate RT. These new approaches also require precise irradiation dose determination for an effective tumor treatment. It is essential now to explore the underlying molecular and cellular mechanisms of response to FLASH irradiation in order to propose appropriate radiotherapeutic treatments for chemo- or radioresistant cells, or, alternatively, combined radiotherapy with immune or chemotherapy strategies for deep-seated tumors. We will review, herein, the last innovative technological approaches in radiation treatments used in cancer radiotherapy, dosimetry challenges for FLASH RT, and *in vitro* cancer cell responses to different types of radiotherapeutic regimens. RT induced alterations in cellular and molecular structures (including proteome modifications) will be described, as well as biological events supporting the emergence of radioresistance. The impact of radiation on tumor microenvironment components will be explored, highlighting the activity switch of immune cell types. Lessons learned from research in different tumor biology areas will be shared with the readers, with emphasis on melanoma and ovarian cancer. We will highlight the differential response of tumor and normal cells exposed to unconventional FLASH irradiation sources. The review is proposing to assess the advantages of using specific cell culture models and to explore a fundamental understanding of FLASH-induced cell perturbations.

Keywords: FLASH radiotherapy, ionizing radiation, laser-plasma accelerators, very high energy electrons, cancer, *in vitro* cell models, tumor microenvironment, radiation-induced cell signaling, proteome, melanoma, ovarian cancer, radioresistance markers

1. Introduction

Based on the ionization mechanism, ionizing radiation is divided into two subcategories: directly and indirectly ionizing [1]. Directly ionizing radiation comprises of charged particles (e.g., electrons, protons, and alpha particles) that deliver energy to the material by Coulomb interaction, while indirectly ionizing radiation consists of electrically neutral particles (e.g., neutrons, gamma rays, and X-rays) that deposit part of their energy in the material in a two-step process: I. A charged particle is released in the form of an electron, neutron, or electron/positron pair and II. The released charged particle interacts electrostatically with the orbital electrons and lose energy by ionizing and exciting atoms along its path. Ionizing radiation produces DNA lesions, such as double-strand breaks (DSBs), which are the main cause of cell death [2].

Recent advances in radiotherapy are pushing the boundaries of cancer treatment, with major innovations aiming to make treatments more precise, efficient, and personalized. However, each advancement faces unique technological challenges. *Proton therapy* is a category of particle therapy that uses high-dose radiation with minimal damage to surrounding healthy tissue [3]. *Carbon ion therapy*, which uses heavier ions, is more effective for some resistant tumors and has a higher biological effect [4]. The main challenges are cost and infrastructure. Particle therapy facilities are still expensive to build and maintain, requiring specialized accelerators and shielding [5]. Patient selection and dosage calculation complexities also need refinement. *Adaptive Radiotherapy (ART)* allows real-time modifications to radiation strategies based on periodic imaging and tumor changes [6, 7]. This personalizes the treatment and minimizes radiation exposure to surrounding tissues. ART requires advanced imaging, automated contouring, and rapid recalculations. The challenge is to integrate these steps into a smooth workflow, with minimized delay and accurate dose delivery. *Radiomics* uses imaging data to analyze tumor characteristics, allowing for more accurate radiation planning [8–10]. Artificial Intelligence (AI) helps predict treatment outcomes, optimize dose distributions, and automate treatment planning [11]. AI models need large, diverse datasets to be accurate and generalizable. Regulatory issues, model transparency, and integration into clinical practice also pose challenges. Radiomics also needs robust standardization to interpret results consistently. New high-dose-rate (HDR) *brachytherapy techniques* allow targeted radiation directly at the tumor site [12]. Image-guided brachytherapy (IGBT) enhances precision by combining imaging with real-time planning [13]. Image registration, particularly in dynamic tissues (e.g., prostate or cervical cancer), can be difficult. The miniaturization of the equipment and reduction of setup time are also critical for its broader adoption. *MRI-Guided Radiotherapy* (MRIgRT) allows for superior soft tissue contrast and real-time imaging [14]. This facilitates more precise targeting of tumors, especially in organs that move (e.g., the liver). The integration of MRI into radiotherapy equipment is technically complex, and MRI-guided systems are expensive. Furthermore, MRI introduces unique dosimetry challenges that need to be addressed for accurate treatment [15]. Combining *radiotherapy* with *immunotherapy* can amplify immune responses against cancer [16, 17]. Low-dose radiation can modulate the immune system, making tumors more susceptible to immune attack. Optimizing dose timing and sequencing to maximize immune activation without enhancing adverse effects remains complex. Additionally, understanding individual immune responses to radiation is challenging, requiring more extensive biomarker research. *FLASH radiotherapy* delivers ultra-high doses in milliseconds instead of minutes [18]. The FLASH-RT is defined by an irradiation delivery at ultrahigh mean dose rates

(>40 Gy s^{-1}) [19]. Preclinical studies show that it kills cancer cells while sparing healthy tissue [20]. Technical challenges include designing machines that reliably produce intense doses and developing dosimetry methods to accurately measure these high-dose, short-duration exposures. Regulatory and clinical validations are also needed. The use of ionizing radiation generated by ultra-intense lasers in FLASH irradiation therapy presents exciting opportunities for cancer treatment [21–25]. Ultra-intense lasers can produce highly focused beams of radiation, allowing to accurately target the tumors, and consequently diminishing damage to adjacent healthy tissues and enhancing the therapeutic ratio. These lasers can deliver exceptionally high dose rates in a very short time, potentially improving tumor control [26]. The rapid delivery may exploit unique biological responses in cancer cells while allowing healthy cells more time to repair. Ionization radiation from lasers can facilitate innovative techniques such as proton or neutron therapy. These methods may offer superior dose distributions and reduced side effects compared to conventional photon therapies. The interaction of ultra-intense laser radiation with biological tissues can induce specific effects, such as enhanced apoptosis (programmed cell death) in cancer cells [25, 27]. This could lead to more effective treatments with fewer sessions [28]. Laser-generated ionization radiation can be combined with other modalities, like immunotherapy or chemotherapy, potentially enhancing overall treatment efficacy by targeting tumors from multiple angles. Ongoing research into the mechanisms of action and optimal delivery methods for laser-based therapies can lead to new protocols and improvements in clinical outcomes [25]. The ionizing radiation from ultra-intense lasers used in FLASH irradiation therapy may offer a promising frontier in oncology, with the potential for cancer treatments with efficacy and less toxicity. Further research is needed to fully realize these opportunities and translate them into clinical practice. These innovations represent significant strides forward but demand extensive cross-disciplinary collaboration, advances in computational power, and considerable financial investment to overcome the technical, logistical, and regulatory hurdles they face.

2. Dosimetry for FLASH radiotherapy

Precise calculation and measurement of radiation dose is a crucial factor for effective tumor treatment by radiotherapy. The absorbed dose is defined as the amount of energy deposited by the radiation in a material mass unit. Equivalent dose is frequently used to assess the extent to which radiation can ionize molecules in irradiated material. It is calculated by multiplying the absorbed dose by a weighting factor that is dependent on the radiation type. It becomes essential to select appropriate measurement tools as they vary with the type of radiation. Charged particles like electrons, protons, or alpha particles can be measured with, for example, ionization chamber, diodes, gafchromic films, or thermoluminescent dosimeter, while uncharged particles like photons and neutrons are measured indirectly by the secondary particles they generate using ionization chamber, calorimeters, neutron dosimeters, and films [29].

2.1 Challenges in dosimetry for FLASH radiotherapy

2.1.1 Response time and saturation

Most standard dosimeters, like ionization chambers and thermoluminescent dosimeters (TLDs), are not designed to measure dose rates above a certain threshold and may saturate or under-respond at FLASH dose rates. These devices typically need a longer response time, making it difficult to capture the true dose delivered over just milliseconds.

2.1.2 Temporal resolution

FLASH therapy demands devices with extremely high temporal resolution to record doses accurately at such short intervals [30]. This is necessary for distinguishing between doses delivered by individual pulses.

2.1.3 Spatial resolution

High spatial resolution is essential to assess dose distribution in the treatment area accurately, especially when targeting tumors close to critical structures [31]. High dose gradients in FLASH make accurate spatial mapping crucial.

2.1.4 Calibration and standardization

The calibration of dosimeters for ultra-high doses in short time spans is challenging [30]. FLASH therapy lacks standardized calibration protocols and reference materials, making it difficult to ensure accuracy and reproducibility across institutions.

2.2 Promising dosimetry approaches for FLASH

2.2.1 Scintillation detectors

Scintillation detectors use materials that emit light when exposed to radiation. They can achieve high temporal resolution and potentially adapt to the high dose rates of FLASH [32]. Plastic scintillators, for instance, are less prone to saturation and can be integrated with high-speed cameras or photomultiplier tubes to capture dose distributions in real time. However, they require corrections for dose-quenching effects and light loss over time.

2.2.2 Radiochromic films

Radiochromic films change color when exposed to radiation, offering a way to visually map dose distributions. They provide high spatial resolution, which helps assess dose gradients, but their slow readout and potential for saturation at ultra-high doses limit their application in real-time FLASH dosimetry [33]. The adjustment of film types and layering techniques could help mitigate some of these issues.

2.2.3 Diamond detectors

Diamond-based detectors have a high atomic number and can provide a stable and fast response, which makes them suitable for high-dose-rate applications [34]. Their stability, radiation hardness, and high charge collection efficiency allow them to capture rapid pulses accurately. However, they are expensive and sensitive to temperature variations, requiring careful handling and calibration.

2.2.4 Cherenkov radiation-based dosimetry

Cherenkov radiation appears once charged particles are faster than the speed of light, emitting a blue glow. This effect can be exploited to create a real-time dosimetry system by capturing the Cherenkov light emitted during FLASH pulses [35]. Fast imaging systems can track these light emissions, providing a direct measure of the dose delivered. However, Cherenkov light can be affected by tissue composition and optical path differences, which require complex calibration.

2.2.5 Fiber-optic dosimeters

Fiber-optic dosimeters use optical fibers that can transmit radiation-induced light emissions to a remote sensor, allowing for high spatial and temporal resolution [36]. They are compact and minimally invasive, making them ideal for real-time measurements. For FLASH, they require specific adaptations to handle ultra-high dose rates without signal saturation or lag.

2.2.6 MicroDiamond and silicon-based detectors

MicroDiamond detectors have high radiation tolerance and are stable under high dose rates [37]. Silicon diodes, which are commonly used in conventional radiotherapy, are being adapted for FLASH with improvements in readout speed and reduction in dose rate dependence. However, like other solid-state devices, they require specific calibration against known standards to address dose rate and linearity challenges.

2.3 New dosimetry approaches

2.3.1 4D dosimetry systems

Techniques combining high-speed imaging with dosimetric measurements, such as high-speed cameras capturing dose distributions in real time, offer a "4D" view (3D spatial + time) of dose distributions [38]. These techniques are being developed to assess both spatial and temporal dose delivery, ensuring that doses are delivered as planned.

2.3.2 Monte Carlo simulations

Monte Carlo methods, traditionally used for radiation dose simulations, are being refined to model FLASH dosimetry [39]. Simulations are adapted to predict dose distributions at ultra-high dose rates, providing a computational means to complement physical dosimetry measurements in cases where direct measurement is challenging.

2.4 Perspectives

To advance FLASH dosimetry, standardizing calibration techniques, developing new materials with faster response rates, and improving real-time imaging will be essential. The combination of physical measurements with computational models, such as Monte Carlo simulations [40], may provide a hybrid solution, ensuring dosimetry accuracy for clinical implementation. Additionally, research institutions and regulatory bodies are working to establish protocols and dose-rate standards specifically suited to FLASH therapy, which would help unify and validate dosimetry approaches across clinics. The development of reliable dosimetry for FLASH is crucial for safe and effective clinical translation, ensuring accurate dose delivery and reproducibility in clinical settings.

2.5 High-energy radiation regimens used in cancer radiotherapy

High energy X-rays are the most common radiations used for cancer radiotherapy (RT) mainly due to their cost efficiency as compared with other radiation sources. Even if they require ultrahigh current parameters to generate FLASH, the feasibility of this type of radiotherapy was proven by Gao and colleagues on animal models [41]. In almost all preclinical and clinical trials for cancer research, FLASH RT was performed either with electrons [19, 41–46], or protons [47–51]. Usually, FLASH RT with electrons and high energy X-rays is exploited for treating superficial cancer lesions, whereas protons are employed for deep-seated tumors [52].

Herein, we have summarized the radiation regimens used in FLASH RT for *in vivo* studies accomplished both on animal models (**Table 1**) and on humans (**Table 2**). FLASH RT has drawn tremendous attention in the cancer domain considering its great potential of sparing the normal tissue toxicity. Pertinent questions were addressed within the scientific community regarding its translation into therapeutic clinical studies [56]. Noteworthy, patient selection was started this year for a randomized phase II clinical study on FLASH RT (dose rate of $2.2–2.7 \times 10^{2}$ Gy/s) compared to CONV RT (dose rate of $3–6 \times 10^{-2}$ Gy/s) for skin cancer treatment [57].

3. Tumor cellular complexity and molecular mechanisms involved in cell response to radiation

3.1 Molecular mechanisms and pathways involved in response to radiotherapy

3.1.1 RTs and genetic material alterations

Genome integrity is continuously challenged by endogenous factors from cellular metabolism (oxidative stress, replication errors) and exogenous agents such as environmental radiation (UV, ionizing), chemicals, or therapeutics. In attempt to preserve their genetic integrity and stability, cells trigger DNA damage repair (DDR) processes/pathways. Unproperly repaired DNA results in cellular senescence, apoptosis, organism susceptibility to immunodeficiency, neurological disorders, and cancer development. To overcome these hurdles, cells respond in a coordinated manner between the detection of DNA damage checkpoints and repair pathways.

The electrons and ions released from molecules exposed to therapeutic ionizing radiations (IRs) (X–/gamma-rays, as well as alpha and beta particles) produce alterations in DNA structure, such as: mismatched bases (MBs), oxidative base damage, single strand breaks (SSBs) or double-strand breaks (DSBs) and activate corresponding repair pathways. The two main repair mechanisms of DNA damage are: homologous recombination (HR) and nonhomologous end-joining (NHEJ) repair [58–60]. IRs produce reactive oxygen or nitrogen species (ROS or RNS), which also induce the formation of a basic (apurinic/apirimidinic) sites in DNA, modifications in sugar moieties, and SSBs [61]. The main consequence of these structural changes is the loss of the clonogenic survival of tumor cells, namely the ability to form clusters.

DSBs are by far the most damaging DNA lesions, which may induce lethality even in situations when they are single and unrepaired. If cells cannot repair DSBs, they can end up with chromosomal aberrations, such as deletions, translocations, and amplifications. Furthermore, these produce altered genes and cell functions, which cause either death or tumor initiation and progression. The two DSB repair pathways,

References	Preclinical study design on animal models	FLASH RT - radiation type - energy - system	FLASH regimen - dose-dose rate	CONV RT -radiation type -energy -system	CONV regimen - dose - dose rate	Key findings
Favaudon [19]	C57BL/6 J mouse models with pulmonary fibrosis (250 animals)	• e^- • 4.5 MeV • Linac, pulsed radiation	• 16–30 Gy • 0.4–0.6 × 10^2 Gy/s	• ^{137}Cs γ-rays • 0.66 MeV • IBL-637 irradiator • constant radiation	• 7.5, 15 or 17 Gy • ≤3 × 10^{-2} Gy/s in single doses	• 17 Gy: CONV: pneumonitis & fibrosis in all animals; no toxic effects with FLASH • 20 Gy FLASH: no fibrogenesis • FLASH: prevention of endothelial cells apoptosis
Favaudon [19]	Swiss mouse models with human breast cancer xenografts (28 animals)	• e^- • 4.5 MeV • Linac, pulsed radiation	• 8.5 Gy • 0.6 × 10^2 Gy/s	• ^{137}Cs γ-rays • 0.66 MeV • IBL-637 irradiator	• 17 Gy • 3.1 × 10^{-2} Gy/s	• FLASH & CONV: similar effects of cancer suppression
Favaudon [19]	Swiss mouse models with human head &neck cancer xenografts (40 animals)	• e^- • 4.5 MeV • Linac, pulsed radiation	• 15, 20, & 25 Gy • 0.6 × 10^2 Gy/s	• X-ray • 200 kV • constant radiation	• 19.5 Gy • 1.2 × 10^{-2} Gy/s	• 25 Gy FLASH: cancer progression stopped @40 days post-RT
Favaudon [19]	C57BL/6 J mouse models with syngeneic luciferase+ lung cancer cells	• e– • 4.5 MeV • Linac	• 15, 23 & 28 Gy • 0.6 × 102 Gy/s	• e– • 4.5 MeV	• 13 & 15 Gy • 3.1 × 10^{-2} Gy/s	• 15 Gy CONV: most with radiation pneumonitis • 28 Gy FLASH: 70% with controlled tumor and none with pneumonitis

References	Preclinical study design on animal models	FLASH RT - radiation type - energy - system	FLASH regimen - dose-dose rate	CONV RT -radiation type -energy -system	CONV regimen - dose - dose rate	Key findings
Vozenin [42]	Goettingen normal pig skin region of 2.6 cm diameter (1 female)	• e– • 4.5 MeV • Kinetron Linac; • e– • 6 MeV • Oriatron eRT6 Linac	• 22–34 Gy • ~3 × 102 Gy/s	• NM	• 22–34 Gy • ~5 Gy/min	• 31 Gy: skin repair with FLASH, but not with CONV @ 15 weeks post-RT • 34 Gy: healthy skin & follicle protection with FLASH & fibrosis without hair reconstruction with CONV @32 weeks post-RT
Vozenin [42]	Phase I dose-escalation study on cats with T2/T3 N0 M0 nasal planum carcinoma (1 female & 5 males)	• e– • 4.5 MeV • Kinetron; • e– • 6 MeV • Oriatron eRT6 Linac	• 25–41 Gy • ~3 × 102 Gy/s • single dose	• NM	• 22–34 Gy • ~5 Gy/min	• Stable depilation within irradiated area, complete tumor response & 84% survival @16 weeks post-RT • Osteonecrosis in 3 animals @ 9–15 months post-FLASH
Fouillade [43]	C57BL/6 J lung mouse models after bilateral thorax irradiation	• e– • 4.5 MeV • Kinetron, chopped mode	• 17 Gy • 1–2.2 × 10^2 mA	• e– • 4.5 MeV • Kinetron • -chopped mode	• 17 Gy • 0.8 × 102 μA	• Reduced cell expansion in resected lungs @ 1-week post-FLASH • Lower consistent DNA foci with FLASH than CONV
Levy [44]	C57BL/6 J mouse models intraperitoneally injected with ovarian cancer cells	• e– • 16 MeV • modified Varian Clinac 21EX [53]	• 16 Gy • 216 Gy/s	• e^- • 16 MeV • Varian Clinac 21EX	• 16 Gy • 0.079 Gy/s	• Post-FLASH: survival time = 90 days & post-CONV: only 10 days • Higher intestinal regeneration with FLASH than CONV @96 hours post-RT

DOI: http://dx.doi.org/10.5772/intechopen.1010060

References	Preclinical study design on animal models	FLASH RT - radiation type - energy - system	FLASH regimen - dose-dose rate	CONV RT -radiation type -energy -system	CONV regimen - dose - dose rate	Key findings
Diffenderfer [47]	C57BL/6 J mouse models with pancreatic cancer	• p+ • 230 MeV	• 12, 18 Gy • 63 Gy/s	• p+ • 230 MeV	• 12, 18 Gy • 0.71, 0.74 Gy/s	• FLASH: no further increase in tumor volume
Konradsson [45]	Dogs with skin cancer: carcinoma, sarcoma, malignant melanoma (10 animals)	• e^- • 10 MeV • Elekta Precise Linac	• 15–35 Gy • 4–5 × 10^2 Gy/s	—	—	• Incomplete or total response in eleven from thirteen irradiated tumors • 35 Gy FLASH: grade 3 side effects to nasal planum
Cunningham [48]	C57BL/6 J mouse models with oral carcinoma cells injected in the leg (12 animals *per* group)	• p+ • 250 MeV • clinical proton pencil beam scanning	• 15 Gy • 57 Gy/s	• p+ • 244 MeV • clinical proton pencil beam scanning	• 15 Gy • 1 Gy/s	• FLASH & CONV: similar tumor volume increase • FLASH: less toxicity
Bley [54]	Randomized phase III study on cats with nasal planum carcinoma - T1-T2, N0 tumors; 7 cats treated with FLASH-radiotherapy and 9 cats treated with standard of care	• e^- • 6 MeV • Oriatron eRT6	• arm 2: 1 × 30 Gy • 15 × 102 Gy/s	• electrons • 6 MeV • Oriatron eRT6	• arm 1: 10 × 4.8 Gy • ~6 Gy/min	• arm 1&2: total recovery @12 months post-RT • post-FLASH: 1 tumor recurrence death @12.1 months & 1 @21.1 months post-CONV • post-FLASH: survival = 656 days—post-CONV: survival = 726 days
Bley [54]	Goettingen normal skin mini pigs with area of 3.5 × 4.5 cm^2 & 8 × 8 cm^2 (2 females)	• e^- • 6 MeV • Oriatron eRT6	• 31 Gy • 150 Gy/s			• 31 Gy: tolerated with low acute effects • harsh skin necrosis @7–9 months post-RT • dermal atrophy @30 months post-RT

References	Preclinical study design on animal models	FLASH RT - radiation type - energy - system	FLASH regimen - dose-dose rate	CONV RT -radiation type -energy -system	CONV regimen - dose - dose rate	Key findings
Gao [41]	BALB/c female mouse models with breast cancer allografts	• e– • 8 MeV • PARTER Linac	• 18 Gy • 10 × 10^2 Gy/s	• e– • 6 MeV • clinical Elekta Precise Linac	• 15 Gy • 10–1 Gy/s	• significant lower tumor volume with FLASH than CONV @63 days post-RT
Gao [41]	BALB/c female mouse models with normal lungs	• high energy X-rays • 6 MeV	• 30 Gy • 7 × 102 Gy/s	• X-rays • 6 MeV	• 30 Gy • 10–1 Gy/s	• survival period = 120 days for FLASH & 86 days for CONV

e^- = electrons; p + = protons; Linac = linear accelerator; @ = after.

Table 1.
Types of radiation used in preclinical studies on animals.

DOI: http://dx.doi.org/10.5772/intechopen.1010060

Reference	Clinical study design on humans	FLASH RT - radiation type - energy - system	FLASH regimen - dose dose - rate	CONV RT - radiation ype - energy - system	CONV regimen - dose - dose rate	Key findings
Bourhis [18]	One patient of 75 years with CD30^{+} T cell T3 N0M0B0 lymphoma	• e– • 5.6 MeV • Oriatron eRT6 Linac	• 15 Gy single dose	• NM	• 110 previous sessions within 10 years with 20–21 Gy • 6–10 fractions	• total wound recovery of 3.5 cm diameter skin tumor @5 months post-RT
Kishan [49, 55]	• MIRAGE phase III randomized study in prostate cancer patients • 77 patients to arm 1 • 79 patients to arm 2	• SBRT arm 2: MRI-guided MRIdian Linac for 2 mm margins	• 40 Gy in 5 fractions of 8 Gy	• SBRT arm 1: CT-guided TrueBeam Linac for 4 mm margins	• 40 Gy in 5 fractions of 8 Gy	• significant minimization of margins & toxic effects in prostate cancer with MRI-guided SRBT than CT-guided SRBT @2 years post-RT
Mascia [50]	• FAST-01 nonrandomized study • 10 patients • 12 bone metastatic sections • 27–81 years old	• p+ • 4.5 MeV • ProBeam Proton Therapy System FDA device	• 15–30 Gy • 0.4–0.6 × 10^2 Gy/s	• photon • 0.66 MeV & • X-rays • 200 kV	• 8 Gy in a single fraction • 0.03 Gy/s	• FLASH: similar pain palliation to CONV @3 months post-RT • total recovery of six metastatic sections • ongoing evaluation every 2 weeks hereafter
Daugherty [51]	• FAST-02 single-arm study • 10 patients • 1–3 bone metastases in the thorax, none in the spine	• p+ • ProBeam Proton Therapy System FDA device	• 8 Gy in a single fraction • ⩾40 Gy/s			• ongoing study for physical suffering & side effects every 6 months on long term post-RT

e^- = electrons; p + = protons; Linac = linear accelerator; @ = after.

Table 2.
Types of radiation used in clinical studies on humans.

NHEJ and HR, are considered genome guardians [62–64]. The cell decision for one or the another is dependent on: DSBs complexity, genetic and genomic background, chromatin state, and CC phase [65]. It implies the participation of multi-protein complexes acting in precise, regulated steps, the so-called "DBS repair arsenal" [63]. Normal cells use both canonical NHEJ (c-NHEJ) as well as HR pathways for resolving DSBs, whereas cancer cells also use error-prone *alternative DNA repair sub pathways.* This is the case when c-NHEJ is impaired by the deletion of essential components to protect and even amplify their genomic instability, oncogenic transformation, and progression [66]. Importantly, in these pathways, there is sequence information loss and an increased risk of unrelated DNA molecule ends joining and creating genomic rearrangements.

It is important to characterize the type of cell death (apoptosis, necrosis, mitotic catastrophe, autophagy, senescence) occurring in tumor cells exposed to newly proposed treatments and its consequences on the modulation of the immune tumor microenvironment. Since the first observation in 1842, multiple forms of regulated cell death in response to different stressors have been described, each based on different molecular machineries [67]. Noteworthy, tumor cellular response to DNA damage induced by IR is dependent on cell genetic profile - specific somatic copy number alterations (CNAs), gene mutations, and gene sets were identified as predictors of radiosensitivity [68]. In addition to nuclear material, any cell type contains a set of common organelles on which they rely to fulfill their metabolic tasks for ensuring tissue and organism homeostasis. These supramolecular-organized cellular constituents have also been acknowledged to be affected by different types of radiation [69, 70]. Among these organelles, the most impacted are the endoplasmic reticulum [71], mitochondria [72, 73], and lysosomes [74]. They play important roles in reducing RT-induced cellular stress by processes such as unfolded protein response, mitophagy, or autophagy.

3.1.2 Nontarget effects of RTs in cancer

In the attempt to achieve a compromise between damaging cancer cells and minimizing damages in adjacent normal tissue, novel advanced technological strategies, such as Intensity-modulated radiation therapy (IMRT), have been developed. This uses irregularly shaped target irradiation volumes from high-dose in the tumor area to low-dose in the vicinity of the tumor with healthy tissue, which, unfortunately, loads the healthy cells with a large volume of low-dose radiation. It is already known that such low doses do not have lethal cell effects, however, they may have unwanted effects in generating even more carcinogenic cells. These data were corroborated with the increasing evidence that cells located either in the vicinity of the irradiation site or away from it are affected and respond by a plethora of events presently known as the nontarget effects of RTs [75–78].

There are three types of such *nontargeted effects of RT* defined based on the relation between the irradiated and nonirradiated cells, as well as on the vicinity to the original site of treatment and accepted by United Nations Scientific Committee of the Effects of Atomic Radiations (UNSCEAR): the bystander effect, the abscopal effect, and the cohort effect [79]. *Bystander effects* are transmitted from irradiated cells to neighboring unirradiated cells within the irradiated volume, leading to biological responses in the recipient cells [80]. *Abscopal* (from the Latin *'ab scopus,'* meaning 'away from the target') *effect* is a systemic effect at a distant metastatic site that is mediated through immunogenic responses [81]. RT-abscopal effects were reported for the first time in melanoma in 1975 [82]. *Cohort effects* occur within the irradiated volume between irradiated cells and are often confused with bystander effects [83].

3.1.3 Proteome and secretome modifications following RT

Quantitative alterations of blood proteome following radiation exposure. The relative expression of proteins can be modulated during cell response to various stress factors or environmental changes [84, 85]. Proteome analysis of serum using MALDI-TOF mass spectrometry for samples obtained from head and neck squamous cell cancer (HNSCC) patients irradiated with low and medium doses of radiation [86] showed significantly distinct profiles as compared to the pre-treatment samples. Approximately 45% of the detected peptide ions revealed modifications of their abundance typically associated with the extent of acute radiation toxicity [86]. Serum proteome analysis of patients with HNSCC or prostate cancer revealed that following RT, plasma apolipoproteins and proteins associated with blood coagulation were predominantly downregulated [87]. Significant changes in the relative expression of 36 serum proteins, mostly associated with lipoproteins, wound-healing, inflammation, and blood coagulation, were reported [88]. Analysis of serum collected from ^{60}Co irradiated nonhuman primates showed that most of the relative changes were associated with proteins participating in the immune system response, platelet degranulation, activation and aggregation, neutrophil degranulation, or protein metabolism [89]. Ex-Rad treatment (a pharmacological agent, candidate for radiation medical countermeasure) reverted only partially the sera proteome modulated by radiation exposure [89]. Similarly, an attenuation of the hematopoietic radiation injury following a 24 h pre-treatment with Gamma-tocotrienol was reported, suggesting a radioprotective effect [90].

Tissue-specific proteome quantitative modifications following radiation exposure. Modifications of the cell and tissue proteome were observed even 2–20 years after RT, as fluctuations in the relative expression of proteins associated with inflammation of neutrophil activity from the mucosal tissue of cancer patients subjected to pelvic radiotherapy [91, 92]. Cardiac tissue of rats subjected to 25 Gy manifested a downregulation of proteins involved in cell adhesion, signal transduction, or GTPase regulation, while proteins associated with ER stress response or endomembrane organization were upregulated during the first 7 days following RT [92]. Moreover, mapping the relative proteome changes of the spleens obtained from mice subjected to 7 Gy gamma radiation revealed Wnt signaling pathway and the actin-cytoskeleton related proteins upregulation [90]. TPM1, TPM4, PCP4, and NPY showed increased expression in the hippocampus of irradiated swine brains [93]. These proteins were previously linked to neurodegenerative diseases [93–95]. Analysis of ex vivo 8 Gy irradiated T lymphocytes, isolated from subjects with radiation-induced breast fibrosis (grade ≥ 2 bf + and grade < 2 bf + patients), revealed ~2000 quantifiable proteins. An increased expression of AK2 in grade ≥ 2 bf + patients and NOX4 increase in PBMCs of both groups were observed [96], suggesting a ROS-related response to the treatment. Increasing levels of statherin and decreasing levels of cystatins, proline-rich proteins, neutrophil defensins, and other inflammatory-related proteins were found in the acquired enamel pellicle of HNSCC patients during and after RT [97]. Moreover, increased expression of alpha-enolase during RT and decreased levels after treatment were revealed in the saliva collected from HNSCC patients before, during, and after RT, while scaffold attachment factor-B1 was upregulated following RT [98]. Hence, RT treatment results in alterations in the relative quantitative expression of proteins from various tissues or body fluids.

3.2 Biological events induced by radiotherapy in tumors: Challenging the "resistance dogma"

The principal rationale for the use of RT in cancer treatment is that tumor cells share common traits, such as inefficient DNA damage repair, altered metabolism, numerous mutations, and a specific, abnormal microenvironment. High-dose radiation severely alters cancer cells, inducing their death and subsequent tumor reduction. In addition, RTs alert immune cells and enhance tumor immune clearance. RT is, however, a double-edged sword treatment because collateral damages are always produced in healthy proliferative cells from bone marrow, epithelial of the gastro-intestinal tract, hair follicles, or in adjacent normal cells. During RT exposure, the risk of occurrence of new neoplastic cells and toxicity increase, and the patient's prognosis worsens. Therefore, the main desiderate in RT is to both eradicate or induce additional susceptibility in tumor cells to other therapies and minimize the negative impact on healthy ones. In addition to direct DNA damage, RT has also a significant impact on tumor microenvironment (TME) components. The radio-resistance dogma has been challenged by improving RT conditions (high dose, high dose *per* fraction, short overall treatment time), assisting them by image guided procedures or combining with other classical therapies. Altogether these approaches successfully overcame insidious resistance mechanisms in an increasing number of cases and improved patients' outcome, in particular of those bearing metastases.

Next, the most documented biological effects of advanced RTs applied in cancer at level of TME components and subcellular structures, as well as their therapeutically potential, will be reviewed, discussed and illustrated in **Figure 1**.

3.2.1 RTs impact on immune cell function and anti-tumor immune response

TME contains a repertoire of cells of the immune system, which have either pro- or anti-tumorigenic functions. Exposing of tumor-associated antigens after RT was reported in colorectal and prostate cancer [99–101], supporting immunogenic cell death. During tumor development, the cancer cells perfect and refine mechanisms to escape immune control and to co-opt immune cells to participate in tumor progression. The discovery of immune checkpoint inhibitors (anti-PD-1/PD-L1 and anti-CTLA4 antibodies), as well as adoptive immune cells (CAR-T), represented major breakthroughs for cancer treatment [102]. However, during radiation exposure, tumors may inhibit the induced immune responses by overexpressing PD-L1 and other suppressive factors of cytotoxic T cell activity, thus minimizing the abscopal effect in clinical applications [103].

The impact of radiation on tumor-associated immune cells will be further briefly reviewed, with emphasis on tumor infiltrating lymphocytes (TILs) and tumor-associated macrophages (TAMs).

3.2.1.1 Impact of radiation on TILs

TILs are critical components of the immune response against tumors, divided into several classes based on their biological functions and phenotype: T lymphocytes, B lymphocytes, and natural killer (NK) cells [104, 105]. The effect of radiation on TILs is complex and context-dependent, determined by a series of factors such as tumor

DOI: http://dx.doi.org/10.5772/intechopen.1010060

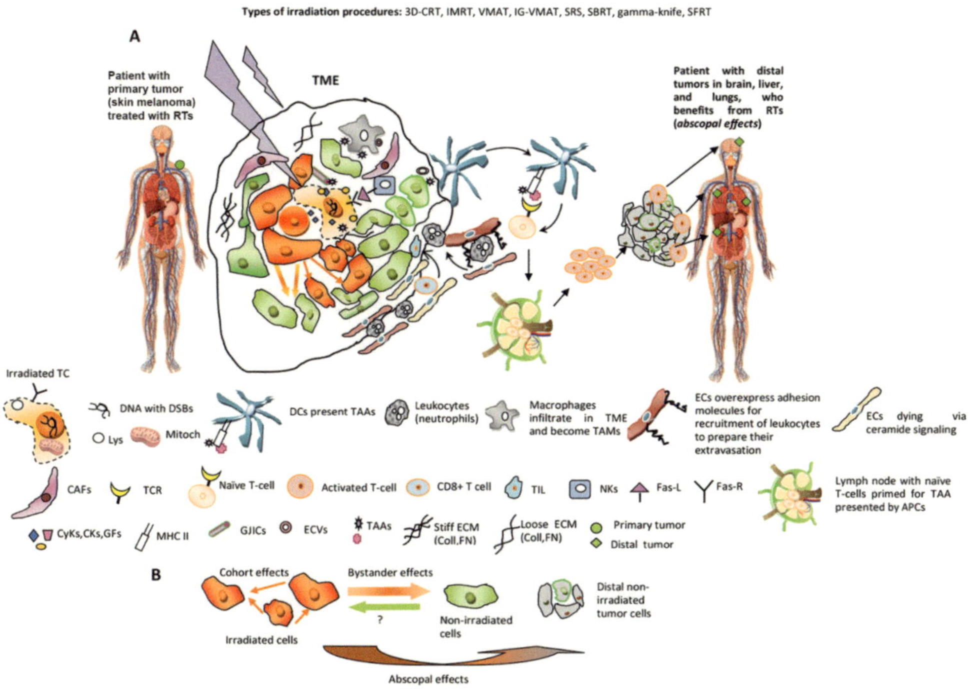

Figure 1.

Tumor microenvironment modulated by radiations. Radiations delivered by different procedures exert on tumors targeted effects (A) and nontargeted effects (B). (A) Irradiated tumors are the site of both anti- and pro-tumorigenic processes/mechanisms. The most significant anti-tumor response is the generation of a "dangerous, pro-inflammatory TME" consisting in: exposure by TCs of damage associated molecular patterns (DAMPs, such as calreticulin and HMB1) or death receptors (Fas-R), release of TAAs, secretion of CKs, CyKs, GFs, extracellular vesicles/exosomes (ECVs), ROS, NOS, NO production, with release of cell content upon cell death due to unrepaired DDBs. The most central cell organelles for integrating and managing the radiation effects are: the endoplasmic reticulum (site of protein synthesis and early processing), mitochondria (cell energy producer and regulator of death processes), and lysosomes (involved in cell residue disposal and death pathway modulator). As an immediate consequence of radiation, cells of the immune system are alerted and mount various anti-tumor responses: enhanced TAA presentation by DCs, more efficiently primed and matured CD8+ T cells, more TILs and NKs are recruited inside the TME, Tregs and MDSCs are reduced, B-cells are increased, TAMs are polarized from M2 (pro-tumorigenic) to M1 (anti-tumorigenic); tumor vasculature express more adhesion molecules to enhance neutrophil extravasation into the TME or is disrupted by endothelial cell death via DNA damage and ceramide signaling; pro-tumorigenic tumor stromal cell (CAFs) contribution (e.g., a stiff ECM) is diminished by reducing matrix stiffness and allowing immune cell penetration and activation. (B) In addition, the non-irradiated cells within the TME situated in the immediate vicinity of irradiated cancer cells and in cell–cell communication via gap-junctions also respond by a plethora of similar processes, known as the bystander effects. It is not clear whether the non-irradiated cells send back signals to the irradiated ones. Moreover, the irradiated tumor cells induce responses in distal nonirradiated tumor cells, apart from the irradiation field (metastatic tumors), by priming T-cells in lymphoid organs to recognize and attack tumors in distal organs, the so-called abscopal effect. Moreover, the irradiated cells communicate between them within the irradiation volume by molecular cross-talks, inducing the cohort effects. One has to keep in mind that radiations also induce unwanted, numerous pro-tumorigenic effects in tumor ecosystems, many of them insufficiently explored or understood (detailed in Section 3). TME-tumor microenvironment; TC-tumor cell; RTs-radiotherapies; Mitoch-mitochondria; Lys-Lysosomes; APCs–antigen presenting cells; DCs-dendritic cells; ECs-endothelial cells; CAFs-cancer associated fibroblast; TIL-tumor infiltrating lymphocytes; NKs-natural killer cells; Fas-R-Fas Receptor; Fas-L-Fas ligand; DSBs-double strand breaks in DNA structure; CyKs-cytokines, CKs-chemokines; GFs-growth factors; MHC-class II- major histocompatibility complex class II; GJICs-Gap-junctions mediated intercellular communications; ECVs-extracellular vesicles; TAAs-tumor associated antigens; Coll-collagen; FN-fibronectin; ROS-reactive oxygen species; NO-nitric oxide; NOS-nitric oxide synthase.

type, irradiation site, radiation dose, fractionation, and contact duration [106–110]. Emerging evidence links higher TIL levels to a better prognosis [111]. Only a higher dose of radiation could induce a decrease in lymphocytes (lymphopenia) [112]. Even more, one of the key factors often cited to explain the superior efficiency and noninvasiveness of FLASH irradiation technology is its ability to significantly reduce damage to circulating lymphocytes. This, in turn, helps preserve the immune system's antitumor activity [113].

3.2.1.1.1 T lymphocytes

Radiation enhances adaptive anti-tumor immunity by boosting antigen presentation. This is achieved by dendritic cells (DCs) and by increasing MHC class I expression on tumor cells, which is crucial for T cells to recognize tumor cells. Some studies on melanoma cells showed a proportional increase of MHC class I with radiation dose [114]. Radiation-induced mutations create a postradiation peptidome with potential neo-epitopes [115], increasing the chances that tumor cells are recognized and eliminated by the adaptive immune system. Radiation promotes T cell migration into the TME through mechanisms like chemokine induction and upregulation of vascular adhesion molecules [106]. FLASH radiation exposure promotes the recruitment of $CD3^+$ T cells into tumors [116]. It induces a higher presence of $CD8^+$ T cells in lung tumors of mice compared to conventional (CONV) RT [117]. Besides its stimulatory effect, in some cases, radiation induces an immunosuppressive effect by: (i) increased programmed death-ligand 1 (PD-L1) and cytotoxic T lymphocyte antigen-4 (CTLA-4) immunosuppressive molecules expression [118]; (ii) TGF-β, which promote the expansion of Tregs and myeloid-derived suppressor cells (MDSCs), thus limiting the effectiveness of the immune response. Moreover, high doses of radiation can also lead to an induced senescence driven by the activation of the tumor suppressors p53 and p21 [119]. Noteworthy, a significant reduction in TGF-β expression and a decrease in fibrosis were observed in tumor-bearing mice [19] and human lung fibroblasts [120] following FLASH radiation exposure. Some studies show increased T cell proliferation after RT without a significant rise in overall TILs [121]. Increased TILs post neoadjuvant chemoradiotherapy in esophageal [122], cervical [123], and rectal cancers [124–126] were also reported. The immune response was higher when receiving a single dose of 15 Gy than in fractionated regimens (5 × 3 Gy) [127]. Tumor rejection was obtained with one dose of 20 Gy in immunocompetent mice in a $CD8^+$ CTLs-dependent fashion. Tregs are more resistant to RT than effector T cells, therefore suppressing the beneficial anti-tumor effects of RT. The irradiation with doses of 10,20, and 30 Gy of patients with cervical cancer decreased $CD8^+$ and $CD4^+$ T cells, while it had no influence on Tregs, installing an immunosuppressive milieu [128]. Oppositely, hypofractionated RT (5 × 8 Gy) reduced the number of Tregs in circulation and within TME [129]. In breast carcinoma, the hypofractionated regimen resulted in abscopal effects when delivered in combination with anti-CTLA-4 IT [130]. The same effects were obtained in non-small cell lung cancer (NSCLC) with hypofractionated stereotactic body radiotherapy (SBRT) [131].

3.2.1.1.2 Natural Killer (NK) cells

NK cells belong to innate lymphoid cells [132] and play a crucial role in the early defense against tumors by targeting MHC-I negative cells that evade T cell recognition [133] and secreting chemokines and cytokines for adaptive immune regulation [134].

NK cells are involved in modulating T cell responses, thereby affecting the adaptive anti-tumor immune response [135]. When activated, NK cells can eradicate tumor cells through different pathways, one of them involving direct release of perforin and granzymes to induce cell lysis [132, 133]. Radiation can have both direct and indirect effects on natural killer (NK) cells, thereby affecting the anti-tumor immune response. Mature NK cells are relatively resistant to radiation, while their precursors are more sensitive. In vitro, RT doses above 30 Gy impair NK cell cytotoxic function before causing cell death or apoptosis [106]. Low doses of radiation enhance NK cell activity, potentially improving their ability to recognize and kill tumor cells [136, 137]. Low-dose IR induces a better expansion of NK cells and cytolytic factors secretion both in vitro and in vivo [138, 139], most probably through the p38 MAPK pathway. In addition to their cytotoxic function, NK cells produce cytokines, particularly interferon-γ (IFN-γ), to modulate adaptive immune responses [140, 141]. Radiation-induced changes in the TME, such as the chemokines and cytokines released, can modulate NK cell recruitment and activation. Conversely, radiation can also induce the expression of immune checkpoint molecules, like PD-L1, which may suppress NK cell activity [142]. The effects of radiation on NK cells have led to a growing interest in combining RT with ITs, such as ICI or cytokine therapy, in order to enhance NK cell-mediated anti-tumor responses [143, 144]. Radiation may also increase the ability of NK cells to target cancer stem cells (CSCs). In vitro, studies demonstrated that radiation (8Gy) upregulated stress markers on tumor cells and increased the frequency of stem-like cells. Combining RT with NK therapy showed improved tumor regression and longer survival [145, 146]. Radiation sensitizes cancer cells to NK-mediated killing, especially when combined with ICI, which boosts NK cell activity [147].

3.2.1.2 Impact of radiation on tumor-associated macrophages (TAMs)

Macrophages (MΦs) residing in TME, known as tumor-associated macrophages (TAMs), play an essential role in both innate and adaptive immunity, being responsible for the phagocytosis and degradation of foreign bodies, as well as in presenting antigens to T cells to initiate immune response. Based on their phenotype, TAMs are classified into pro-inflammatory, cancer cells killing M1- MΦs and M2-polarized MΦs promoting cancer progression and metastasis. TAMs predominantly exhibit a status similar to the M2 phenotype, which is associated with tumor progression, aggressiveness, and poor prognostic.

Given their major role in cancer, several ways for targeting TAMs were proposed, ranging from the impairment of MΦs accumulation in TME or depleting TAMs to repolarization from M2 to M1 phenotype [148–151]. TAMs are also key resistance factors developed during CT, IT, and RT, resulting in treatment failure and local relapse. The mechanisms involve DDR, hypoxic TME, soluble factors induced by radiation, secretion of exosomes, and tumor metabolic reprogramming [152, 153].

RT acts either as immune stimulating factor or activates immunosuppressive TAMs, influencing their recruitment and phenotype depending on the irradiation dose [154, 155]. Thus, high doses (>10 Gy) increase M2-like TAMs accumulation leading to T cell suppressive response; moderate doses (1–10 Gy) promote the M1-phenotype; low doses have no effect on M2 to M1 polarization. Different signaling pathways are involved, such as DNA damage induction, NF-kB activation, ROS generation, and MAPK phosphorylation [156]. There are different approaches to boost RT anti-cancer effect, such as inhibition of bone marrow derived TAMs influx in irradiated tumors [157] or driving TAMs populations within TME toward an M1 phenotype [158, 159]. In this context, several innovative RT modalities, including SBRT, heavy-particle-beam therapy, proton-beam RT, carbon ion therapy, and FLASH RT, have been reported [160–163]. Of these, ultrahigh dose rate (UHDR) FLASH RT became a promising therapeutic strategy in cancer due to its ability to protect healthy tissues while keeping its tumoricidal capacity [164, 165]. By delivering a high dose of radiation in a very short time (<1 s), FLASH RT can modulate MΦs behavior (**Table 3**) by inducing cell death and triggering enhanced immune responses,

System type (References)	Treatment radiation strategy (total dose/dose rate)	Key findings
Normal whole-brain [166]	10 Gy FLASH-X-rays (37 Gy/s with dose-rate in the slice of 12×10^3 Gy/s) or CONV (0.05 Gy/s)	↓ reactive astrogliosis.
Normal whole-brain [167]	10 Gy FLASH (>100 Gy/s) or CONV-X-ray (0.07–0.1 Gy/s)	↓ activation of astrogliosis and microglia
Normal whole-brain [168]	10 Gy FLASH ($5.6x 10^6$ Gy/s) or COVN-X-rays (0.1 Gy/s)	limitation of reactive gliosis
Normal whole-brain [169]	30 Gy FLASH (200 or 300 Gy/s) *vs.* CONV (0,13Gy/s)	↓ microglial activation (CD68); ↓ pro-inflammatory cytokines (*i.e.*, IL-6, IL-1β)
Normal whole-brain [170]	8 Gy FLASH (7.7×10^3 Gy/s) or CONV (4.4×10^6 Gy/s).	↓microglial activation
Normal whole-brain [164]	3 × 10 Gy FLASH (5.6×10^6 Gy/s) or CONV (0.09 Gy/s)	↓ microglia cells in the hippocampus and in the medial prefrontal cortex
Normal whole-brain [171]	10 Gy FLASH (120 Gy/s) and standard dose rate (SDR) Bragg Peak Proton Irradiation (0.17 Gy/s)	↓ microglial activation, ↓ neuroinflammation
Lung normal (thorax) [19]	17 Gy FLASH (60 Gy/s) vs. CONV γ-rays (0.03 Gy/s)	MΦ infiltration in alveolar lumens
Lung normal (thorax) [43]	17 Gy in a microsecond pulse or in a limited number of pulses of 1–2 Gy each in 100 ms temporal sequence	↓ *Tgfb1* expression in interstitial MΦs
Whole abdominal [172]	15 Gy or 10 Gy FLASH- X-ray (>150 Gy/s, instantaneous dose rate of $>5.5 \times 10^5$ Gy/s) or CONV	↓ MΦs ($F4/80^+$) compared with control group
Whole abdominal (intestine) [173]	14 Gy FLASHpr (125.3 ± 4.8 Gy/s) or CONVpr (0.82 ± 0.1 Gy/s)	↑ MΦ infiltration, ↑ *Tgfb1* expression and TGF-β secretion
Lewis lung carcinoma [174]	22 Gy + 4 × 0.5 Gy compared with 24 Gy alone	↓ CD206 TAMs expression, ↓ IL-10-secreting TAMs, ↑ anti-tumor TAMs (M1)
Highly metastatic breast cancer [174]	3 × 20 Gy on primary tumor +4 × 0.5 Gy whole lung irradiation	↑ splenic MΦs activation
Lewis lung carcinoma [117]	15 Gy FLASH (352.1 Gy/s) vs. CONV (0.060 Gy/s)	no effect on MLC activation in MΦs
Lewis lung carcinoma tumor model [175]	18 Gy FLASHpr (>60 Gy/s) vs. CONVpr (<0.05 Gy/s)	↓ pro-tumorigenic M2-like TAMs in tumors; ↑ infiltration of anti-tumor M1-like TAMs

System type (References)	Treatment radiation strategy (total dose/dose rate)	Key findings
Murine melanoma model [176]	MRT of 401.23 Gy (11,704 Gy/s) or temporally fractionated MRT 3 × 133.41 Gy (12,870 Gy/s)	↑ melanin-containing CD68$^+$ MΦ cells
Murine models of ovarian cancer [177]	14 Gy FLASH (210 Gy/s) or CONV (0.126 Gy/s), +/− αPD-1 immunotherapy	↓ M2/M1 TAMs ratios only in combination with αPD-1 immunotherapy
Breast cancer [178]	10 Gy abdominal X-ray FLASH (125 Gy/s) vs. CONV (0.2 Gy/s)	↓ TAMs (F4/80$^+$) infiltration
Glioma [179]	25 Gy FLASHpr (257 ± 2 Gy/s) or CONVpr (4 ± 0.02 Gy/s).	↓ microglia (CD45^{low} CD11b/c$^+$ cells) density; CD8$^+$ TAMs not increased
Diffuse Midline Glioma (DMG) [180]	15 Gy FLASH (90 Gy/s) vs. CONV (2 Gy/min)	↓ type 1 interferon alpha receptor (IFNAR+) in microglial cells; ↓ IFN1 pathway in non-resident MΦs
Lung adenocarcinoma and glioblastoma [165]	20 Gy or (3 × 8 or 2 × 6) Gy FLASH (≥2000 Gy/s) vs. CONV (0.1 Gy/s) ±	↑ MΦ s (CD11b$^+$ F4/80$^+$) Effective in an extreme immunodeficient context

Table 3.
Summary of in vivo *studies on the FLASH effect on normal or tumor-associated macrophages (TAMs) grouped by type of cancer.*

anti-tumor functions, and phagocytic ability. Moreover, by targeting TAMs, FLASH can act synergistically with CT or IT, thus increasing cancer therapy efficacy. A few examples highlighting the advantages of FLASH RT are described below.

The first proof-of-concept experiment that demonstrated the neuroprotective effects of FLASH proton (FLASHpr) irradiation *in vivo* was reported by Dokic et al. in 2022. SDR irradiation triggered a significant 1.75-fold increase of CD68$^+$ cells compared to FLASHpr [171]. Whole-abdomen mouse irradiation with a single-dose FLASHpr resulted in a significantly greater MΦ infiltration into the regenerating intestines, which corresponded to reduced normal intestinal tissue toxicity [173]. Subcutaneously implanted 3LL cells were exposed to CONV IR or FLASH IR. FLASH did not induce MLC activation in MΦs, and the number of p-MLC$^+$/F4/80$^+$ cells was similar to that observed in CONV-IR or nonirradiated animals. Vascular collapse in tumors after CONV-IR and not FLASH-IR could be linked with increased p-MLC and independent of MΦ infiltration [117]. In an orthotopic mouse model of NSCLC, mice bearing lung tumors were treated with a single fraction of proton RT, with either CONVpr or FLASHpr delivered bilaterally [175]. The authors demonstrated that both irradiation regimens decreased significantly the number of F4/80$^+$ MΦs in the TME. Also, FLASHpr treatment increased the CTLs to MΦs ratio within the lung tumors. FLASH irradiation resulted in increased CD8$^+$ T cell trafficking into the tumor, probably as a result of the reduced number of M2-like MΦs at the tumor site [175]. A different approach was employed for melanoma treatment using temporally fractionated microbeam radiotherapy (MRT) intersected at 45 degrees angle over 3 days [176]. The obtained data suggest that fractionated UHDR MRT increases MΦ infiltration at the tumor site leading to tumor remission and reduced metastasis [176]. No differences were observed in murine models of ovarian cancer (OC) within the ID8 ovarian TIME regarding the frequency of M2/M1 when comparing non-irradiated with high dose CONV- or FLASH-irradiated mouse omenta. Both CONV-IR and FLASH-IR in combination with αPD-1 IT reduced

M2/M1 MΦ ratios in the ovarian TIME of ID8 tumor model. Similar results were seen in the UPK10 tumor model. The authors highlight the beneficial effect of FLASH-RT on both αPD-1 resistant and sensitive in vivo models of OC and the overall potential of UHDR irradiation in reducing immunosuppressive characteristics of the TIME and improving PD-1/PD-L1 immune checkpoint blockade in OC [177].

Recently, the contribution of the immune response to the anti-tumor efficacy of FLASH using murine and human tumor models (lung adenocarcinoma and glioblastoma) has been investigated [165]. Regarding systemic immune responses, a similar pattern between FLASH- and CONV-irradiated mice was revealed. Overall, both FLASH and CONV, or their combination with anti-CTLA4, lead to comparable local and systemic tumor responses. However, unlike CONV, FLASH induces an anti-tumor response in severely immunocompromised conditions and thus could represent a useful therapeutic approach for cold tumors or severely immunocompromised patients [165].

3.2.2 RTs and impact on tumor vasculature and stroma

Tumor blood vessels are characterized by irregular morphology and increased permeability. Immature blood vessels and endothelial cells in the tumor are more sensitive to radiation [181]. At doses higher than 5–10 Gy the vasculature is damaged [182]. Low-dose radiation schedules as well as SRS or SBRT are producing angiogenesis and neovascularization and endothelial cell survival [183, 184]. In contrast, radiation doses higher than 10 Gy/fraction are consistent with severe damage of tumor vascular network and TME [184].

Cancer-associated fibroblasts (CAFs) represent a heterogeneous cellular group with multiple roles in tumor growth and ECM remodeling [185]. CAFs are targets in combined therapies to overcome the resistance developed by ICI [186]. CAFs, as well as normal fibroblasts (FBs), are radioresistant to high doses of radiation (50 Gy). When they are exposed to doses higher than 10 Gy, CAFs undergo an irreversible process of senescence. Conversely, when applying low doses, these induce reversible DNA damage and exit from CC arrest. Senescent FBs release enzymes, cytokines, GFs, and ROS, thereby building an environment favorable to tumor formation. Stromal FBs are recruited into TME following radiation exposure, which triggers an inflammatory response followed by fibrosis. This process creates a stiff and proreactive stroma, further promoting tumor growth [187]. Increasing the dose/fraction reduces matrix stiffness and FBs differentiation and activation, which is favorable to tumor progression. High-dose RT in some cancers may reduce pro-tumorigenic CAFs activity [188] by changing their secretome [189].

4. Radiotherapy in melanoma

Melanoma is a type of tumor associated mainly with skin, the cutaneous melanoma (CM). Despite the fact that CM represents only 4–7% of skin cancers, 80% of deaths are caused by this neoplasm. Due to an insidious evolution, aggressiveness, and remarkable intrinsic or acquired resistance to various therapies, CM has an increasing incidence worldwide, affecting both sexes (more alarming in young people under 30 years of age), and still remaining a serious public health issue. The main incriminating risk factor for the development of CMs remains the environmental UV radiation.

Melanoma was excluded for a long period from RTs ultimately due to lack of overall survival benefits, due to insufficient knowledge about the relationship between total dose, dose per fraction, duration of therapy and site of treatment. The expansion of modern concepts of RT coupled with imagistics, such as computer tomography, magnetic resonance (MRI), and positron tomography (PET) together with molecular targeted personalized treatments emerged as novel options in the battle against melanoma, which is still one of the most deadly types of cancer [76, 190–193].

4.1 The latest advanced technological RT strategies applied to melanoma treatment

The approved therapies in melanoma include surgery, CTs, ITs, RTs and combinations thereof [193–195]. They are all continuously updated, based on accumulated knowledge about mutational landscape, novel molecular key players, and pathways, which operate specifically in this type of neoplasm, as well as in relation with the immune system.

RT is used in melanoma treatment: (a) as *alternative to surgery* [196–199]; (b) as *adjuvant or postoperative therapy* [200, 201]; (c) as *palliative therapy* [202, 203]; (d) as *enhancer of other therapies*-combinatorials with ITs and CTs using different type of techniques, such as: external beam RT, interventional RT, proton therapy, and different types of inhibitors (e.g., ICI), are increasing immunostimulatory potential of radiation and show promising results in term of survival outcomes [204–207]. Another combination is RT and *hyperthermia* (HT), which relies on both radio-sensitization, as well as direct hyperthermic cytotoxicity against radio-resistant tumor cells, and it is less toxic than other conventional therapies [208–211].

There are two types of RTs: External beam RT (EBRT) and Internal RT (IRT). In EBRT, different types of beams produced outside the patient target the body tumor site. EBRT includes multiple types of procedures: a) *3D-conformal RT (3D-CRT)*, in which the shape of the applied irradiation is limited to simple squares or rectangles due to uniform intensity across each radiation field; b) *Intensity modulated radiation therapy (IMRT)* is a type of 3D-CRT, which uses many smaller beams than 3D-CRT. When IMRT is combined with image-guidance (IG) technologies, the approach is called IG-IMRT [212]. This combinatorial approach is more suitable to precisely deliver the radiation than other forms of radiation therapy; c) *Volumetric modulated arc therapy (VMAT)* delivers the radiation dose as the treatment machine rotates and adjusts the radiation dose to the tumor geometry. When it is coupled with IG it is called IG-VMAT; d) *SRS* allows the delivery of a single high dose RT directly to target cancer tissues, thereby eliminating the tumor with few side effects. In the case of melanoma brain metastasis, the late studies recommend a multidisciplinary approach, including surgery, SRS, systemic immuno, and targeted chemo-therapy [213, 214]. Gamma knife radiosurgery is a type of SRS using gamma (γ) rays suitable for therapy of small tumors in head, neck, or brain areas [215]. The name comes from the use of very precise, directed beams to target the tumors and produce low damage within the normal tissues. Future studies using variable SRS doses and larger cohorts are needed in order to validate the protocols. SBRT is suitable to address tumors throughout the body, excluding the brain and spine [216, 217].

In IRT, as opposed to EBRT, the source of radiations is inside the patient and it can be i) solid (*brachytherapy/BRT*), when the radioactive material is placed inside the body directly or near tumor, as an implant, which may come as tube, wire, capsule, seeds, pellets, or ii) liquid (*systemic radiotherapy/SRT*), when the treatment is delivered by swallowing, intravenously (IV) or by injection, enters the blood stream, and reaches tissues throughout entire body, targeting and killing cancer cells. BRT represents an option for patients with bleeding or pain or for those with tumors difficult to be excised. *Spatially fractionated radiation therapy (SFRT)* is a novel alternative, still under testing, consisting of high and low doses throughout the tumor in one or more fractions. The distribution of radiation throughout tumor in standard therapy is in homogenous low-doses fractions, whereas SFRT uses a lethal dose (peak) in one part of the tumor, while in other portions sublethal (valley) doses [218]. The rationale was to spare as much as possible from the normal tissue being damaged. In traditional therapy, a large number of small fractions of low dose (25–30 fractions of 2 Gy) are used, as opposed to spatial fractionation, which uses a large dose in peak areas (20 Gy) and a reduced one in the valleys (3 Gy). The bystander effects resulting from SFRT can be beneficial in normal tissue, whereas tumor eradication occurs with this type of treatment; although not fully understood, it is under current molecular investigation in order to develop new therapeutic strategies [219].

4.2 RTs broaden and streamline traditional therapeutic approaches in melanoma management

RT is now embedded in melanoma multidisciplinary therapeutic approaches aiming to prolong cancer remission, increase overall survival, and manage more efficiently the metastatic stages. RT is also aligned with the concept of "precision medicine" [220]. To obtain extrinsic tumor radiosensitization, multiple parameters should be considered, such as high dose *per* fraction, high dose rate (to induce a saturation of tumor DNA repair mechanisms), short overall treatment time, radiosensitizers, and high LET. Procedures involving RT are adjusted to the melanoma type and patient clinical stage; modern models based on spatial and temporal modulation of radiation dose on response have been elaborated [77]. The progress made in deciphering pathways induced in tumors by nontargeted effects (the bystander and abscopal effects) has allowed the development of novel, complex therapeutic strategies known as chemo-immuno-radio-therapies (CIRTs), considering the most promising solutions of each type [221].

The cancer stage benefiting the most from the nontargeted effects of RT (mainly the abscopal effect) is the metastatic one. There are numerous recruiting clinical trials of *IRTs* in melanoma involving combinatorials of RTs approaches and classical targeted ITs: (a) Ipilimumab (Iplmab) (anti-CTLA-4) and different RTs: NCT02097732, NCT 01449279, NCT01970527, NCT01996202), NCT02303990; Tremelimumab (anti-CTLA-4) and hypofractionated RT: NCT02639026. All these examples clearly demonstrate that RT optimize significantly anti-tumor response induced by ITs [222]; (b) RT and anti-PD-1 antibody (MK-3475) (run by Sidney Kimmel Cancer Center at Thomas Jefferson University, Philadelphia, Pennsylvania, US); (c) vaccines based on DCs loaded with autologous tumor lysate combined with RT and/or pre-leukapheresis IFN-α [223]. Chemoradiotherapies (*CRTs*) that combine the benefits of chemotherapeutics and radiation therapy in cancer eradication have not been employed in melanoma so far to our knowledge.

5. Radiotherapy in ovarian cancer

5.1 Perspectives on past and future radiotherapeutic regimes

ICI therapy of advanced OC has shown only modest benefits in the management of this disease due to a low response rate to treatment [224–230]. Research results in this area point to a highly immunosuppressive environment as the cause for the low success rate of anti-cancer immunotherapy [231]. This, together with low mutational burden in OC, explains the lack of visibility of OC tumors by the immune system. The performed clinical trials showed, however, promising tolerability to IT agents [229]. Thus, the development of safe and effective combination therapies is urgently needed in order to increase response to this type of treatment. In some cancers, hypofractionated (high-dose) radiotherapy (HDRT; >5Gy per fraction) has been shown to enhance the anti-tumor immune response [127, 232–235] and synergize with PD-1/PD-L1 blockade [115, 236]. However, CONV-RT has limited use in OC due to increased associated abdominopelvic toxicity [237–239]. CONV fractionated RT uses mostly 2–3 Gy doses to induce cancer cell death. When doses higher than 8 Gy/fraction are used, the vascular system is impacted through endothelial cell apoptosis [240]. Recently, George Coukos' Lab tested a new strategy of applying low-dose CONV radiotherapy (LDRT; 0.5 or 1 Gy every 2 weeks; total radiation dose 6 or 13 Gy, respectively) to reverse tumor immune desertification and resistance to ICI of OC patients enrolled in the RACIN phase I clinical trial [241]. Patients responded by regression of metastatic solid tumors when LDRT was combined with orthogonal IT. Biopsies confirmed the pro-inflammatory effect of LDRT at both tested doses and the simultaneous activation of innate and adaptive immunity, which was associated with tumor response. However, the persistence of side effects from the immunomodulatory cocktail was higher than expected and led to treatment discontinuation.

A recent preclinical study at Stanford University indicated that abdominopelvic FLASH irradiation enhances response to anti-PD1 IT in mouse models of OC [177]. These effects are mediated *via* increased T cell infiltration and activation within the *omentum* TME, as well as decreased Tregs-to-effector ratio in the tumor and the mesenteric lymph nodes. Radiation-induced intestinal injury was decreased as compared to CONV irradiation conditions. Further, equal abscopal response [242] and systemic anti-tumor immunity were induced by FLASH and CONV X-ray when integrated with ICI.

5.2 Tissue transglutaminase (TG2) as a targetable marker of survival to radiotherapy in different types of cancer

Tissue transglutaminase (TG2) is a multifunctional protein that plays important roles in cancer cells resistance to CT and RT. In this regard, TG2 operates *via* activating the NF-κB survival pathway or by promoting the signaling cascade initiated by TG2-regulated cell adhesion to the extracellular matrix [243]. TG2 is upregulated in glioblastoma [244], ovarian [245], pancreatic [246], lung [247], and breast cancer [248]. TG2 overexpression was associated with an unfavorable prognosis of patients with OC [249], lung cancer [250], and pancreatic cancer [251]. TG2 is involved in resistance to RT in lung cancer, melanoma, glioblastoma, and laryngeal cancer. In lung cancer, TG2 was shown to participate in the nuclear DDR by interacting with Topo IIα through the TG-ase domain, upon irradiation with a 365 nm-pulsed nitrogen laser [252]. Artificially increasing TG2's biosynthesis did not influence melanoma cells proliferation and clonal expansion after irradiation [253]. TG2 knockout (KO) has reduced the proliferation, clonal expansion, and spheroid formation capacity of irradiated cells, meaning TG2 expression can be correlated with radioresistance. TG2 and SDC1 were found to be overexpressed in radioresistant glioblastoma multiforme (GBM) [254]. SDC1-TG2 complex regulates GBM radioresistance by fusing autophagosomes with lysosomes. Irradiated astrocytes secrete TG2 in order to stimulate a pro-tumorigenic TME [255]. Astrocytes survived in mouse brains after irradiation, subsequently causing injected glioma cells to form more aggressive tumors. Astrocytes accomplished this by increased secretion of TG2 in the ECM, which persists even after tumor regression. This further promotes the formation of new tumors and enhances stemness and survival of cancer cells. In a glioblastoma orthotopic xenograft mouse model, radiation-induced TG2 regulated mesenchymal transdifferentiation that supports cancer cell therapy resistance; TGM2 gene silencing or TG2 inhibitor GK921 interfered with this phenotype and prolonged mice survival [256]. Increased TG2 gene copy number was detected in esophageal cancer, where it was correlated with nodal metastasis and tumor invasion [257]. TG2 levels are further increased in chemo- or radio-resistant cancer cells [258]. In laryngeal cancer, the expression of TG2 and BNIP3 affect postoperative RT response [259]. High TG2 and/or low BNIP3 tumors were correlated with negative clinical outcomes for patients undergoing postoperative radiation therapy.

When secreted, TG2 forms bridges to connect integrin receptors and fibronectin (FN) [260], thereby increasing FN bundling [261] and cells capacity to adhere to FN-rich tissues [260]. This favors tumor growth by increasing cancer cells proliferation and invasion abilities and by increasing resistance to anti-cancer treatments, including RT. We and others [262–265] have participated in the effort of developing molecules that would prevent TG2-FN interaction in order to minimize its contribution to different pathologies. TG2-FN inhibitors are able to sensitize OC cells to CT drugs such as paclitaxel [264] and represent a promising strategy to combat tumor progression in cancer lesions highly expressing TG2. NC9 or AA9 inhibitors of TG2 transamidation activity [266] were shown to have modulatory activity on cancer cell invasion and metabolism [267]. Overall, TG2 targeting shows moderate apoptosis-inducing properties but has increased potential as sensitizing agent to other therapeutics, including RT. TG2-FN irreversible inhibitor KCC009 induces radiosenzitization in lung adenocarcinoma cells, irrespective of p53 status [268]. Importantly, TG2 expression in the host tissues is also implicated in modulating the T cell-mediated anti-tumor immune response [269, 270]. In the absence of TG2 in the host, there is a decreased ascites accumulation in the peritoneal cavity and a lower number of OC cells in ascites [269]. $CD8^+$ T cells from the TME of tumor-bearing TG2KO mice have a predominant effector memory phenotype and increased cancer-killing capacity. In human OC tumors, high expression of TG2 in the stroma is correlated with decreased $CD8^+$ T cell infiltration into tumor tissue, which makes TG2 an important immunomodulatory target. Moreover, Wang et al. [270] have shown in three different solid tumors that TG2 facilitates the formation of covalent CXCL12–keratin-19 (KRT19) heterodimers, which assemble into filamentous networks. This restricts the access of $CD8^+$ T cells into tumors. The CXCL12-KRT19 coating leads to IT resistance, which could be reversed by either TG2 or KRT19 elimination from PDA cells. Therefore, a continuation of the efforts to develop potent and selective TG2 inhibitors and/or their combinatorial use with other therapies shows high potential of translation into the clinic.

6. Conclusions

The progress in radiotherapy (RT) meets nowadays technological challenges and major innovations for precise, efficient and personalized treatments. These challenges and innovations are based on extensive cross-disciplinary collaborations, advances in computational power, and considerable financial investment to overcome the technical and logistical difficulties.

With this chapter, we gave an overview of some innovative technologies and critical methodologies that may have potential for effective clinical translation, with an emphasis on FLASH RT using ionizing radiation. Some challenges were defined in Sections 1 and 2, such as the use of ionizing radiation generated by ultra-intense lasers in FLASH irradiation therapy for cancer treatment, and the delivery of accurate doses and reproducibility in FLASH dosimetry and standardizing calibration techniques. Further, *in vitro* protocols are essential to be developed in order to understand cellular behavior when exposed to FLASH irradiation. In this context, it is known that the ionizing radiation results in significant changes in the tumor microenvironment (TME) while also representing an immunological adjuvant. In Section 3, we discussed the aspects of tumor cellular complexity and molecular mechanisms involved in cell response to radiation. From a biological point of view, it is essential to explore molecular and cellular mechanisms behind the cellular response to FLASH irradiation in order to propose appropriate RT treatments for radioresistant cells, such as melanoma cells, or combined RT with immunotherapy (IT) or chemotherapy (CT) strategies for deep-seated tumors, such as ovarian cancer (OC). In Section 4, we presented the cellular and molecular responses to radiotherapy in melanoma microenvironment while in Section 5 the cellular and molecular responses in OC.

The scientific impact of understanding the FLASH irradiation effect on cancer cells may potentially bring progress in new tissue culture technical approaches relevant for the exploration of fundamental molecular aspects of the induced FLASH effect. It can also support the development of new strategies to advance the stability and uniformity of new FLASH irradiation sources. Molecular *in vitro* investigations will represent valuable assets to associate different mechanisms involved in resistance to environmental and/or therapeutic stressors and for therapeutic combinations.

Acknowledgements

This research was supported by Institute of Atomic Physics (IFA) through the ELI-RO_2024_10 Project, No. 10/2024; and the Romanian Ministry of Education and Research under the Romanian National Nucleus Program LAPLAS VII under Contract No. 30 N/2023. We acknowledge the support of Project No. 1/2024 of the Research Program of the Institute of Biochemistry of the Romanian Academy. We acknowledge the authors of the article entitled "First *in vitro* cell co-culture experiments using laser-induced high-energy electron FLASH irradiation for the development of anti-cancer therapeutic strategies" in which the use of the *in vitro* protocol with cell co-cultures exposed to FLASH high energy was first described [25]. We acknowledge the support of National Interest Infrastructure facility IOSIN–CETAL at INFLPR.

Conflict of interest

The authors declare no conflict of interest.

Author details

Livia Sima[1], Monica Tudor[1], Stefana Orobeti[1,2], Anca Filimon[1], Gabriela Negroiu[1], Cristian Munteanu[1], Gabriela Chiritoiu[1], Anca Roseanu[1], Paula Florian[1], Madalina Icriverzi[1], Ioana Porosnicu[2], Daniel Avram[2], Alexandra Bran[2], Sandel Simion[2], Ion Tiseanu[2] and Felix Sima[2*]

1 Institute of Biochemistry of the Romanian Academy, Bucharest, Romania

2 National Institute for Laser, Plasma and Radiation Physics, Magurele, Romania

*Address all correspondence to: felix.sima@inflpr.ro

References

[1] United Nations Scientific Committee on the Effects of Atomic Radiation. Ionizing Radiation: Sources and Biological Effects. Report to the General Assembly with Scientific Annexes; 1982

[2] Vignard J, Mirey G, Salles B. Ionizing-radiation induced DNA double-strand breaks: A direct and indirect lighting up. Radiotherapy and Oncology. 2013;**108**:362-369

[3] Khan FM, Gibbons JP. Khan's the Physics of Radiation Therapy. Lippincott Williams & Wilkins; 2014

[4] Haettner E, Iwase H, Schardt D. Experimental fragmentation studies with 12C therapy beams. Radiation Protection Dosimetry. 2006;**122**:485-487

[5] Li Y, Li X, Yang J, Wang S, Tang M, Xia J, et al. Flourish of proton and carbon ion radiotherapy in China. Frontiers in Oncology. 2022;**12**

[6] Glide-Hurst CK, Lee P, Yock AD, Olsen JR, Cao M, Siddiqui F, et al. Adaptive radiation therapy (ART) strategies and technical considerations: A state of the ART review from NRG oncology. International Journal of Radiation Oncology, Biology, Physics. 2021;**109**:1054-1075

[7] Sonke JJ, Aznar M, Rasch C. Adaptive radiotherapy for anatomical changes. Seminars in Radiation Oncology. 2019;**29**:245-257

[8] Lambin P, Leijenaar RTH, Deist TM, Peerlings J, De Jong EEC, Van Timmeren J, et al. Radiomics: The bridge between medical imaging and personalized medicine. Nature Reviews. Clinical Oncology. 2017;**14**:749-762

[9] Avanzo M, Stancanello J, El Naqa I. Beyond imaging: The promise of radiomics. Physica Medica. 2017;**38**:122-139

[10] van Timmeren JE, Cester D, Tanadini-Lang S, Alkadhi H, Baessler B. Radiomics in medical imaging-"how-to" guide and critical reflection. Insights Into Imaging. 2020;**11**

[11] Bera K, Braman N, Gupta A, Velcheti V, Madabhushi A. Predicting cancer outcomes with radiomics and artificial intelligence in radiology. Nature Reviews Clinical Oncology. 2021;**19**:132-146

[12] Martinez AA, Demanes J, Vargas C, Schour L, Ghilezan M, Gustafson GS. High-dose-rate prostate brachytherapy: An excellent accelerated-hypofractionated treatment for favorable prostate cancer. American Journal of Clinical Oncology. 2010;**33**:481-488

[13] Sturdza A, Pötter R, Fokdal LU, Haie-Meder C, Tan LT, Mazeron R, et al. Image guided brachytherapy in locally advanced cervical cancer: Improved pelvic control and survival in RetroEMBRACE, a multicenter cohort study. Radiotherapy and Oncology. 2016;**120**:428-433

[14] Raaymakers BW, Jürgenliemk-Schulz IM, Bol GH, Glitzner M, Kotte ANTJ, Van Asselen B, et al. First patients treated with a 1.5 T MRI-Linac: Clinical proof of concept of a high-precision, high-field MRI guided radiotherapy treatment. Physics in Medicine and Biology. 2017;**62**:L41-L50

[15] Low DA, Moran JM, Dempsey JF, Dong L, Oldham M. Dosimetry tools and

techniques for IMRT. Medical Physics. 2011;**38**:1313-1338

[16] Formenti SC, Demaria S. Combining radiotherapy and cancer immunotherapy: A paradigm shift. Journal of the National Cancer Institute. 2013;**105**:256-265

[17] Demaria S, Bhardwaj N, McBride WH, Formenti SC. Combining radiotherapy and immunotherapy: A revived partnership. International Journal of Radiation Oncology, Biology, Physics. 2005;**63**:655-666

[18] Bourhis J, Sozzi WJ, Jorge PG, Gaide O, Bailat C, Duclos F, et al. Treatment of a first patient with FLASH-radiotherapy. Radiotherapy and Oncology. 2019;**139**:18-22

[19] Favaudon V, Caplier L, Monceau V, Pouzoulet F, Sayarath M, Fouillade C, et al. Ultrahigh dose-rate FLASH irradiation increases the differential response between normal and tumor tissue in mice. Science Translational Medicine. 2014;**6**:245-293

[20] Bourhis J, Montay-Gruel P, Gonçalves Jorge P, Bailat C, Petit B, Ollivier J, et al. Clinical translation of FLASH radiotherapy: Why and how? Radiotherapy and Oncology. 2019;**139**:11-17

[21] Lundh O, Lim J, Rechatin C, Ammoura L, Ben-Ismaïl A, Davoine X, et al. Few femtosecond, few kiloampere electron bunch produced by a laser–plasma accelerator. Nature Physics. 2011;**7**:219-222

[22] Faure J, Rechatin C, Norlin A, Lifschitz A, Glinec Y, Malka V. Controlled injection and acceleration of electrons in plasma wakefields by colliding laser pulses. Nature. 2006;**444**:737-739

[23] Bin J, Obst-Huebl L, Mao JH, Nakamura K, Geulig LD, Chang H, et al. A new platform for ultra-high dose rate radiobiological research using the BELLA PW laser proton beamline. Scientific Reports. 2022;**12**:1-15

[24] Chaudhary P, Milluzzo G, Ahmed H, Odlozilik B, McMurray A, Prise KM, et al. Radiobiology experiments with ultra-high dose rate laser-driven protons: Methodology and state-of-the-art. Frontiers of Physics. 2021;**9**:624963

[25] Orobeti S, Sima LE, Porosnicu I, Diplasu C, Giubega G, Cojocaru G, et al. First in vitro cell co-culture experiments using laser-induced high-energy electron FLASH irradiation for the development of anti-cancer therapeutic strategies. Scientific Reports. 2024;**14**:1-13

[26] Fuchs T, Szymanowski H, Oelfke U, Glinec Y, Rechatin C, Faure J, et al. Treatment planning for laser-accelerated very-high energy electrons. Physics in Medicine and Biology. 2009;**54**:3315

[27] Rigaud O, Fortunel NO, Vaigot P, Cadio E, Martin MT, Lundh O, et al. Exploring ultrashort high-energy electron-induced damage in human carcinoma cells. Cell Death & Disease. 2010;**1**:e73-e73

[28] Laschinsky L, Baumann M, Beyreuther E, Enghardt W, Kaluza M, Karsch L, et al. Radiobiological effectiveness of laser accelerated electrons in comparison to electron beams from a conventional linear accelerator. Journal of Radiation Research. 2012;**53**:395-403

[29] Podgorsak EB. Radiation Oncology Physics: A Handbook for Teachers and Students. International Atomic Energy Agency; 2005

[30] Favaudon V, Lentz J-M, Heinrich S, Patriarca A, de Marzi L, Fouillade C,

et al. Time-resolved dosimetry of pulsed electron beams in very high dose-rate, FLASH irradiation for radiotherapy preclinical studies. Nuclear Instruments and Methods in Physics Research A. 2019;**944**:162537

[31] Andreo P. On the clinical spatial resolution achievable with protons and heavier charged particle radiotherapy beams. Physics in Medicine and Biology. 2009;**54**:N205

[32] Beaulieu L, Beddar S. Review of plastic and liquid scintillation dosimetry for photon, electron, and proton therapy. Physics in Medicine and Biology. 2016;**61**:R305

[33] Villoing D, Koumeir C, Bongrand A, Guertin A, Haddad F, Métivier V, et al. Technical note: Proton beam dosimetry at ultra-high dose rates (FLASH): Evaluation of GAFchromic™ (EBT3, EBT-XD) and OrthoChromic (OC-1) film performances. Medical Physics. 2022;**49**:2732-2745

[34] Pettinato S, Salvatori S. Diamond-based detection systems for tomorrow's precision dosimetry. Nuclear Instruments and Methods in Physics Research A. 2024;**1059**:168974

[35] Ashraf MR, Rahman M, Zhang R, Williams BB, Gladstone DJ, Pogue BW, et al. Dosimetry for FLASH radiotherapy: A review of tools and the role of radioluminescence and cherenkov emission. Frontiers of Physics. 2020;**8**:570033

[36] Fontbonne JM, Iltis G, Ban G, Battala A, Vernhes JC, Tillier J, et al. Scintillating fiber dosimeter for radiation therapy accelerator. IEEE Transactions on Nuclear Science. 2002;**49 I**:2223-2227

[37] Ahmad UF, Wudil YS, Imam A, Isa NF, Gondal MA, Al-Osta MA. Applications of carbon-based diamond detectors: A critical review. Materials Today Communications. 2023;**36**:106409

[38] Kozicki M, Maras P. On the measurement of radiation isocenter for medical accelerators using 3D polymer gel dosimetry. Introduction, application, and good practices. Measurement. 2023;**221**:113585

[39] Romano F, Bailat C, Jorge PG, Lerch MLF, Darafsheh A. Ultra-high dose rate dosimetry: Challenges and opportunities for FLASH radiation therapy. Medical Physics. 2022;**49**:4912-4932

[40] Zlateva Y, El Naqa I. Cherenkov emission dosimetry for electron beam radiotherapy: A Monte Carlo feasibility study of absolute dose prediction. IFMBE Proceedings. 2015;**51**:828-831

[41] Gao F, Yang Y, Zhu H, Wang J, Xiao D, Zhou Z, et al. First demonstration of the FLASH effect with ultrahigh dose rate high-energy X-rays. Radiotherapy and Oncology. 2022;**166**:44-50

[42] Vozenin MC, De Fornel P, Petersson K, Favaudon V, Jaccard M, Germond JF, et al. The advantage of FLASH radiotherapy confirmed in mini-pig and cat-cancer patients. Clinical Cancer Research. 2019;**25**:35-42

[43] Fouillade C, Curras-Alonso S, Giuranno L, Quelennec E, Heinrich S, Bonnet-Boissinot S, et al. FLASH irradiation spares lung progenitor cells and limits the incidence of radio-induced senescence. Clinical Cancer Research. 2020;**26**:1497-1506

[44] Levy K, Natarajan S, Wang J, Chow S, Eggold JT, Loo PE, et al. Abdominal FLASH irradiation reduces radiation-induced gastrointestinal

toxicity for the treatment of ovarian cancer in mice. Scientific Reports. 2020;**10**:1-14

[45] Konradsson E, Arendt ML, Bastholm Jensen K, Børresen B, Hansen AE, Bäck S, et al. Establishment and initial experience of clinical FLASH radiotherapy in canine cancer patients. Frontiers in Oncology. 2021;**11**:658004

[46] Li P, Hu M, Liu M, Ren X, Liu D, Liu J, et al. The efficacy and safety of different systemic combination therapies on advanced hepatocellular carcinoma: A systematic review and meta-analysis. Frontiers in Oncology. 2023;**13**:1197782

[47] Diffenderfer ES, Verginadis II, Kim MM, Shoniyozov K, Velalopoulou A, Goia D, et al. Design, implementation, and *in vivo* validation of a novel proton FLASH radiation therapy system. International Journal of Radiation Oncology, Biology, Physics. 2020;**106**:440-448

[48] Cunningham S, McCauley S, Vairamani K, Speth J, Girdhani S, Abel E, et al. FLASH proton pencil beam scanning irradiation minimizes radiation-induced leg contracture and skin toxicity in mice. Cancers (Basel). 2021;**13**:1-15

[49] Kishan AU, Ma TM, Lamb JM, Casado M, Wilhalme H, Low DA, et al. Magnetic resonance imaging-guided vs computed tomography-guided stereotactic body radiotherapy for prostate cancer: The MIRAGE randomized clinical trial. JAMA Oncology. 2023;**9**:365-373

[50] Mascia AE, Daugherty EC, Zhang Y, Lee E, Xiao Z, Sertorio M, et al. Proton FLASH radiotherapy for the treatment of symptomatic bone metastases: The FAST-01 nonrandomized trial. JAMA Oncology. 2023;**9**:62-69

[51] Daugherty EC, Zhang Y, Xiao Z, Mascia AE, Sertorio M, Woo J, et al. FLASH radiotherapy for the treatment of symptomatic bone metastases in the thorax (FAST-02): Protocol for a prospective study of a novel radiotherapy approach. Radiation Oncology. 2024;**19**:1-9

[52] Atkinson J, Bezak E, Le H, Kempson I. The current status of FLASH particle therapy: A systematic review. Physical and Engineering Sciences in Medicine. 2023;**46**:529-560

[53] Schüler E, Trovati S, King G, Lartey F, Rafat M, Villegas M, et al. Experimental platform for ultra-high dose rate FLASH irradiation of small animals using a clinical linear accelerator. International Journal of Radiation Oncology, Biology, Physics. 2017;**97**:195-203

[54] Bley CR, Wolf F, Jorge PG, Grilj V, Petridis I, Petit B, et al. Dose- and volume-limiting late toxicity of FLASH radiotherapy in cats with squamous cell carcinoma of the nasal planum and in mini pigs. Clinical Cancer Research. 2022;**28**:3814-3823

[55] Kishan AU, Lamb JM, Wilhalme H, Casado M, Chong N, Zello L, et al. Magnetic resonance imaging versus computed tomography guidance for stereotactic body radiotherapy in prostate cancer: 2-year outcomes from the MIRAGE randomized clinical trial. European Urology. 2024

[56] Loo BW, Verginadis II, Sørensen BS, Mascia AE, Perentesis JP, Koong AC, et al. Navigating the critical translational questions for implementing FLASH in the clinic. Seminars in Radiation Oncology. 2024;**34**:351-364

[57] Kinj R, Gaide O, Jeanneret-Sozzi W, Dafni U, Viguet-Carrin S, Sagittario E, et al. Randomized phase II selection trial

of FLASH and conventional radiotherapy for patients with localized cutaneous squamous cell carcinoma or basal cell carcinoma: A study protocol. Clinical and Translational Radiation Oncology. 2024;**45**:100743

[58] Lomax ME, Folkes LK, O'Neill P. Biological consequences of radiation-induced DNA damage: Relevance to radiotherapy. Clinical Oncology (Royal College of Radiologists). 2013;**25**:578-585

[59] Sutherland BM, Bennett PV, Sidorkina O, Laval J. Clustered DNA damages induced in isolated DNA and in human cells by low doses of ionizing radiation. Proceedings of the National Academy of Sciences of the United States of America. 2000;**97**:103-108

[60] Sutherland BM, Bennett PV, Sutherland JC, Laval J. Clustered DNA damages induced by X rays in human cells. Radiation Research. 2002;**157**(6):611-616

[61] Redon CE, Nakamura AJ, Zhang YW, Ji J, Bonner WM, Kinders RJ, et al. Histone gammaH2AX and poly(ADP-ribose) as clinical pharmacodynamic biomarkers. Clinical Cancer Research. 2010;**16**:4532-4542

[62] Mao Z, Bozzella M, Seluanov A, Gorbunova V. Comparison of nonhomologous end joining and homologous recombination in human cells. DNA Repair (Amst). 2008;**7**:1765-1771

[63] Mladenov E, Magin S, Soni A, Iliakis G. DNA double-strand break repair as determinant of cellular radiosensitivity to killing and target in radiation therapy. Frontiers in Oncology. 2013;**3**

[64] Trenner A, Sartori AA. Harnessing DNA double-strand break repair for cancer treatment. Frontiers in Oncology. 2019;**9**

[65] Brandsma I, Gent DC. Pathway choice in DNA double strand break repair: Observations of a balancing act. Genome Integrity. 2012;**3**

[66] Caracciolo D, Riillo C, Di Martino MT, Tagliaferri P, Tassone P. Alternative non-homologous end-joining: Error-prone DNA repair as cancer's Achilles' heel. Cancers (Basel). 2021;**13**:1-14

[67] Tang D, Kang R, Vanden BT, Vandenabeele P, Kroemer G. The molecular machinery of regulated cell death. Cell Research. 2019;**29**:347-364

[68] Yard BD, Adams DJ, Chie EK, Tamayo P, Battaglia JS, Gopal P, et al. A genetic basis for the variation in the vulnerability of cancer to DNA damage. Nature Communications. 2016;**7**:1-14

[69] Somosy Z. Radiation response of cell organelles. Micron. 2000;**31**:165-181

[70] Song WJ, Juan WH, Li QH. Biological effects of radiation on cancer cells. Military Medical Research. 2018;**5**:20

[71] Sabouny R, Fraunberger E, Geoffrion M, Ng ACH, Baird SD, Screaton RA, et al. The Keap1-Nrf2 stress response pathway promotes mitochondrial hyperfusion through degradation of the mitochondrial fission protein Drp1. Antioxidants & Redox Signaling. 2017;**27**:1447-1459

[72] Au WH, Miller-Fleming L, Sanchez-Martinez A, Lee JAK, Twyning MJ, Prag HA, et al. Activation of the Keap1/Nrf2 pathway suppresses mitochondrial dysfunction, oxidative stress, and motor phenotypes in C9orf72 ALS/FTD models. Life Science Alliance. 2024:7

[73] Averbeck D, Rodriguez-Lafrasse C. Role of mitochondria in radiation responses: Epigenetic, metabolic, and signaling impacts. International Journal of Molecular Sciences. 2021;**22**

[74] Zhang X, Wang J, Li X, Wang D. Lysosomes contribute to radioresistance in cancer. Cancer Letters. 2018;**439**:39-46

[75] Daguenet E, Louati S, Wozny AS, Vial N, Gras M, Guy JB, et al. Radiation-induced bystander and abscopal effects: Important lessons from preclinical models. British Journal of Cancer. 2020;**123**:339-348

[76] Espenel S, Vallard A, Rancoule C, Garcia MA, Guy JB, Chargari C, et al. Melanoma: Last call for radiotherapy. Critical Reviews in Oncology/ Hematology. 2017;**110**:13-19

[77] Suchowerska N, Ebert MA, McKenzie DR, Jackson M. A review of in vitro experimental evidence for the effect of spatial and temporal modulation of radiation dose on response. Acta Oncologica. 2010;**49**:1344-1353

[78] McMahon SJ, Butterworth KT, Trainor C, McGarry CK, O'Sullivan JM, Schettino G, et al. A kinetic-based model of radiation-induced intercellular signalling. PLoS One. 2013;**8**

[79] Blyth BJ, Sykes PJ. Radiation-induced bystander effects: What are they, and how relevant are they to human radiation exposures? Radiation Research. 2011;**176**:139-157

[80] Azzam EI, De Toledo SM, Little JB. Direct evidence for the participation of gap junction-mediated intercellular communication in the transmission of damage signals from α-particle irradiated to nonirradiated cells. Proceedings of the National Academy of Sciences of the United States of America. 2001;**98**:473-478

[81] Nabrinsky E, Macklis J, Bitran J. A review of the abscopal effect in the era of immunotherapy. Cureus. 2022

[82] Kingsley DPE. An interesting case of possible abscopal effect in malignant melanoma. The British Journal of Radiology. 1975;**48**:863-866

[83] Wang R, Zhou T, Liu W, Zuo L. Molecular mechanism of bystander effects and related abscopal/cohort effects in cancer therapy. Oncotarget. 2018;**9**:18637-18647

[84] Munteanu CVA, Chiritoiu GN, Chiritoiu M, Ghenea S, Petrescu AJ, Petrescu SM. Affinity proteomics and deglycoproteomics uncover novel EDEM2 endogenous substrates and an integrative ERAD network. Molecular & Cellular Proteomics. 2021;**20**

[85] Munteanu CVA, Chirițoiu GN, Petrescu AJ, Petrescu ȘM. Defining the altered glycoproteomic space of the early secretory pathway by class I mannosidase pharmacological inhibition. Frontiers in Molecular Biosciences. 2023;**9**

[86] Widłak P, Pietrowska M, Polańska J, Rutkowski T, Jelonek K, Kalinowska-Herok M, et al. Radiotherapy-related changes in serum proteome patterns of head and neck cancer patients; the effect of low and medium doses of radiation delivered to large volumes of normal tissue. Journal of Translational Medicine. 2013;**11**

[87] Widlak P, Jelonek K, Wojakowska A, Pietrowska M, Polanska J, Marczak Ł, et al. Serum proteome signature of radiation response: Upregulation of inflammation-related factors and downregulation of apolipoproteins and coagulation factors in cancer patients

treated with radiation therapy--a pilot study. International Journal of Radiation Oncology, Biology, Physics. 2015;**92**:1108-1115

[88] Ouerhani A, Chiappetta G, Souiai O, Mahjoubi H, Vinh J. Investigation of serum proteome homeostasis during radiation therapy by a quantitative proteomics approach. Bioscience Reports. 2019;**39**

[89] Carpenter AD, Li Y, Janocha BL, Wise SY, Fatanmi OO, Maniar M, et al. Analysis of the proteomic profile in serum of irradiated nonhuman primates treated with ex-rad, a radiation medical countermeasure. Journal of Proteome Research. 2023;**22**:1116-1126

[90] Cheema AK, Byrum SD, Sharma NK, Altadill T, Kumar VP, Biswas S, et al. Proteomic changes in mouse spleen after radiation-induced injury and its modulation by gamma-tocotrienol. Radiation Research. 2018;**190**:449-463

[91] Devarakonda S, Thorsell A, Hedenström P, Rezapour A, Heden L, Banerjee S, et al. Low-grade intestinal inflammation two decades after pelvic radiotherapy. eBioMedicine. 2023;**94**

[92] Kim BH, Jung JW, Han D, Cha MJ, Chang JH. One-week dynamic changes in cardiac proteomes after cardiac radioablation in experimental rat model. Frontiers in Cardiovascular Medicine. 2022;**9**:898222

[93] Iacono D, Hatch K, Murphy EK, Post J, Cole RN, Perl DP, et al. Proteomic changes in the hippocampus of large mammals after total-body low dose radiation. PLoS One. 2024;**19**

[94] Wang X, Williams D, Müller I, Lemieux M, Dukart R, Maia IBL, et al. Tau interactome analyses in CRISPR-Cas9 engineered neuronal cells reveal ATPase-dependent binding of wild-type but not P301L tau to non-muscle myosins. Scientific Reports. 2019;**9**

[95] Brettle M, Patel S, Fath T. Tropomyosins in the healthy and diseased nervous system. Brain Research Bulletin. 2016;**126**:311-323

[96] Lacombe J, Brengues M, Mangé A, Bourgier C, Gourgou S, Pèlegrin A, et al. Quantitative proteomic analysis reveals AK2 as potential biomarker for late normal tissue radiotoxicity. Radiation Oncology. 2019;**14**

[97] Ventura TMO, Ribeiro NR, Taira EA, de Souza-e-Silva CM, Rubira CMF, da Santos PSS, et al. Radiotherapy changes acquired enamel pellicle proteome in head and neck cancer patients. Journal of Dentistry. 2021:108

[98] Ventura TMO, Ribeiro NR, Taira EA, de Lima LA, Dionizio A, Rubira CMF, et al. Radiotherapy changes the salivary proteome in head and neck cancer patients: Evaluation before, during, and after treatment. Clinical Oral Investigations. 2022;**26**:225-258

[99] Nesslinger NJ, Sahota RA, Stone B, Johnson K, Chima N, King C, et al. Standard treatments induce antigen-specific immune responses in prostate cancer. Clinical Cancer Research. 2007;**13**:1493-1502

[100] Schaue D, Comin-Anduix B, Ribas A, Zhang L, Goodglick L, Sayre JW, et al. T-cell responses to survivin in cancer patients undergoing radiation therapy. Clinical Cancer Research. 2008;**14**:4883-4890

[101] Gulley JL, Arlen PM, Bastian A, Morin S, Marte J, Beetham P, et al. Combining a recombinant cancer vaccine

with standard definitive radiotherapy in patients with localized prostate cancer. Clinical Cancer Research. 2005;**11**:3353-3362

[102] Lei X, Lei Y, Li JK, Du WX, Li RG, Yang J, et al. Immune cells within the tumor microenvironment: Biological functions and roles in cancer immunotherapy. Cancer Letters. 2020;**470**:126-133

[103] Topalian SL, Taube JM, Anders RA, Pardoll DM. Mechanism-driven biomarkers to guide immune checkpoint blockade in cancer therapy. Nature Reviews. Cancer. 2016;**16**:275-287

[104] Dunn GP, Old LJ, Schreiber RD. The three Es of cancer immunoediting. Annual Review of Immunology. 2004;**22**:329-360

[105] Fridman WH, Pagès F, Sauts-Fridman C, Galon J. The immune contexture in human tumours: Impact on clinical outcome. Nature Reviews Cancer. 2012;**12**:298-306

[106] Monjazeb AM, Schalper KA, Villarroel-Espindola F, Nguyen A, Shiao SL, Young K. Effects of radiation on the tumor microenvironment. Seminars in Radiation Oncology. 2020;**30**:145

[107] Park B, Yee C, Lee KM. The effect of radiation on the immune response to cancers. International Journal of Molecular Sciences. 2014;**15**:927-943

[108] Kaur P, Asea A. Radiation-induced effects and the immune system in cancer. Frontiers in Oncology. 2012;**2**

[109] Brandmaier A, Formenti SC. The impact of radiation therapy on innate and adaptive tumor immunity. Seminars in Radiation Oncology. 2020;**30**:139-144

[110] Guo S, Yao Y, Tang Y, Xin Z, Wu D, Ni C, et al. Radiation-induced tumor immune microenvironments and potential targets for combination therapy. Signal Transduction and Targeted Therapy. 2023;**8**:1-22

[111] Lanitis E, Dangaj D, Irving M, Coukos G. Mechanisms regulating T-cell infiltration and activity in solid tumors. Annals of Oncology. 2017;**28**:xii18-xii32

[112] Takanen S, Bottero M, Nisticò P, Sanguineti G. A systematic review on the impact of hypofractionated and stereotactic radiotherapy on immune cell subpopulations in cancer patients. Cancers (Basel). 2022;**14**

[113] Jin JY, Gu A, Wang W, Oleinick NL, Machtay M, Kong FM. Ultra-high dose rate effect on circulating immune cells: A potential mechanism for FLASH effect? Radiotherapy and Oncology. 2020;**149**:55-62

[114] Reits EA, Hodge JW, Herberts CA, Groothuis TA, Chakraborty M, Wansley EK, et al. Radiation modulates the peptide repertoire, enhances MHC class I expression, and induces successful antitumor immunotherapy. The Journal of Experimental Medicine. 2006;**203**:1259-1271

[115] Formenti SC, Rudqvist NP, Golden E, Cooper B, Wennerberg E, Lhuillier C, et al. Radiotherapy induces responses of lung cancer to CTLA-4 blockade. Nature Medicine. 2018;**24**:1845-1851

[116] Rama N, Saha T, Shukla S, Goda C, Milewski D, Mascia AE, et al. Improved tumor control through T-cell infiltration modulated by ultra-high dose rate proton FLASH using a clinical pencil beam scanning proton system. International Journal of

Radiation Oncology*Biology*Physics. 2019;**105**:S164-S165

[117] Kim YE, Gwak SH, Hong BJ, Oh JM, Choi HS, Kim MS, et al. Effects of ultra-high doserate FLASH irradiation on the tumor microenvironment in Lewis lung carcinoma: Role of myosin light chain. International Journal of Radiation Oncology, Biology, Physics. 2021;**109**:1440-1453

[118] Formenti SC, Demaria S. Systemic effects of local radiotherapy. The Lancet Oncology. 2009;**10**:718

[119] Jiao Y, Cao F, Liu H. Radiation-induced cell death and its mechanisms. Health Physics. 2022;**123**:376

[120] Buonanno M, Grilj V, Brenner DJ. Biological effects in normal cells exposed to FLASH dose rate protons. Radiotherapy and Oncology. 2019;**139**:51-55

[121] Suwa T, Saio M, Umemura N, Yamashita T, Toida M, Shibata T, et al. Preoperative radiotherapy contributes to induction of proliferative activity of CD8+ tumor-infiltrating T-cells in oral squamous cell carcinoma. Oncology Reports. 2006;**15**:757-763

[122] Kelly RJ, Zaidi AH, Smith MA, Omstead AN, Kosovec JE, Matsui D, et al. The dynamic and transient immune microenvironment in locally advanced esophageal adenocarcinoma post chemoradiation. Annals of Surgery. 2018;**268**:992-999

[123] Dorta-Estremera S, Colbert LE, Nookala SS, Yanamandra AV, Yang G, Delgado A, et al. Kinetics of intratumoral immune cell activation during chemoradiation for cervical cancer. International Journal of Radiation Oncology, Biology, Physics. 2018;**102**:593-600

[124] Lim YJ, Koh J, Kim S, Jeon SR, Chie EK, Kim K, et al. Chemoradiation-induced alteration of programmed death-ligand 1 and CD8+ tumor-infiltrating lymphocytes identified patients with poor prognosis in rectal cancer: A matched comparison analysis. International Journal of Radiation Oncology, Biology, Physics. 2017;**99**:1216-1224

[125] Teng F, Meng X, Kong L, Mu D, Zhu H, Liu S, et al. Tumor-infiltrating lymphocytes, forkhead box P3, programmed death ligand-1, and cytotoxic T lymphocyte-associated antigen-4 expressions before and after neoadjuvant chemoradiation in rectal cancer. Translational Research. 2015;**166**:721-732.e1

[126] Lim S, Chua W, Cheng C, Descallar J, Ng W, Solomon M, et al. Effect of neoadjuvant chemoradiation on tumor-infiltrating/associated lymphocytes in locally advanced rectal cancers. Anticancer Research. 2014;**34**:6505-6513

[127] Lugade AA, Moran JP, Gerber SA, Rose RC, Frelinger JG, Lord EM. Local radiation therapy of B16 melanoma tumors increases the generation of tumor antigen-specific effector cells that traffic to the tumor. Journal of Immunology. 2005;**174**:7516-7523

[128] Liang Y, Lü W, Zhang X, Lü B. Tumor-infiltrating CD8+ and FOXP3+ lymphocytes before and after neoadjuvant chemotherapy in cervical cancer. Diagnostic Pathology. 2018;**13**

[129] Wei S, Egenti MU, Teitz-Tennenbaum S, Zou W, Chang AE. Effects of tumor irradiation on host T-regulatory cells and systemic immunity in the context of adoptive T-cell therapy in mice. Journal of Immunotherapy. 2013;**36**:124-132

[130] Dewan MZ, Galloway AE, Kawashima N, Dewyngaert JK, Babb JS, Formenti SC, et al. Fractionated but not single-dose radiotherapy induces an immune-mediated abscopal effect when combined with anti-CTLA-4 antibody. Clinical Cancer Research. 2009;**15**:5379-5388

[131] Rutkowski J, Ślebioda T, Kmieć Z, Zaucha R. Changes in systemic immune response after stereotactic ablative radiotherapy. Preliminary results of a prospective study in patients with early lung cancer. Polish Archives of Internal Medicine. 2017;**127**:245-253

[132] Artis D, Spits H. The biology of innate lymphoid cells. Nature. 2015;**517**:293-301

[133] Lodoen MB, Lanier LL. Natural killer cells as an initial defense against pathogens. Current Opinion in Immunology. 2006;**18**:391

[134] Mamessier E, Sylvain A, Bertucci F, Castellano R, Finetti P, Houvenaeghel G, et al. Human breast tumor cells induce self-tolerance mechanisms to avoid NKG2D-mediated and DNAM-mediated NK cell recognition. Cancer Research. 2011;**71**:6621-6632

[135] Mora-Bitria L, Asquith B. Germline natural killer cell receptors modulating the T cell response. Frontiers in Immunology. 2024;**15**:1477991

[136] Lacoste-Collin L, Jozan S, Cances-Lauwers V, Pipy B, Gasset G, Caratero C, et al. Effect of continuous irradiation with a very low dose of gamma rays on life span and the immune system in SJL mice prone to B-cell lymphoma. Radiation Research. 2007;**168**:725-732

[137] Ina Y, Sakai K. Activation of immunological network by chronic low-dose-rate irradiation in wild-type mouse strains: Analysis of immune cell populations and surface molecules. International Journal of Radiation Biology. 2005;**81**:721-729

[138] Cheda A, Wrembel-Wargocka J, Lisiak E, Nowosielska EM, Marciniak M, Janiak MK. Single low doses of X rays inhibit the development of experimental tumor metastases and trigger the activities of NK cells in mice. Radiation Research. 2004;**161**:335-340

[139] Hashimoto S, Shirato H, Hosokawa M, Nishioka T, Kuramitsu Y, Matushita K, et al. The suppression of metastases and the change in host immune response after low-dose total-body irradiation in tumor-bearing rats. Radiation Research. 1999;**151**:717-724

[140] Vivier E, Raulet DH, Moretta A, Caligiuri MA, Zitvogel L, Lanier LL, et al. Innate or adaptive immunity? The example of natural killer cells. Science. 2011;**331**:44-49

[141] Luetke-Eversloh M, Cicek BB, Siracusa F, Thom JT, Hamann A, Frischbutter S, et al. NK cells gain higher IFN-γ competence during terminal differentiation. European Journal of Immunology. 2014;**44**:2074-2084

[142] Shen MJ, Xu LJ, Yang L, Tsai Y, Keng PC, Chen Y, et al. Radiation alters PD-L1/NKG2D ligand levels in lung cancer cells and leads to immune escape from NK cell cytotoxicity via IL-6-MEK/Erk signaling pathway. Oncotarget. 2017;**8**:80506-80520

[143] Zheng W, Ling S, Cao Y, Shao C, Sun X. Combined use of NK cells and radiotherapy in the treatment of solid tumors. Frontiers in Immunology. 2023;**14**:1306534

[144] Fujisaki H, Kakuda H, Shimasaki N, Imai C, Ma J, Lockey T, et al. Expansion

of highly cytotoxic human natural killer cells for cancer cell therapy. Cancer Research. 2009;**69**:4010-4017

[145] Ames E, Canter RJ, Grossenbacher SK, Mac S, Smith RC, Monjazeb AM, et al. Enhanced targeting of stem-like solid tumor cells with radiation and natural killer cells. Oncoimmunology. 2015;**4**:1-11

[146] Kim KW, Jeong JUK, Lee KH, Uong TNT, Rhee JH, Ahn SJ, et al. Combined NK cell therapy and radiation therapy exhibit long-term therapeutic and antimetastatic effects in a human triple negative breast cancer model. International Journal of Radiation Oncology, Biology, Physics. 2020;**108**:115-125

[147] Makowska A, Lelabi N, Nothbaum C, Shen L, Busson P, Tran TTB, et al. Radiotherapy combined with PD-1 inhibition increases NK cell cytotoxicity towards nasopharyngeal carcinoma cells. Cells. 2021;**10**

[148] Cassetta L, Pollard JW. Targeting macrophages: Therapeutic approaches in cancer. Nature Reviews Drug Discovery. 2018;**17**:887-904

[149] Lin Y, Xu J, Lan H. Tumor-associated macrophages in tumor metastasis: Biological roles and clinical therapeutic applications. Journal of Hematology & Oncology. 2019;**12**

[150] Pathria P, Louis TL, Varner JA. Targeting tumor-associated macrophages in cancer. Trends in Immunology. 2019;**40**:310-327

[151] Li C, Xu X, Wei S, Jiang P, Xue L, Wang J. Tumor-associated macrophages: Potential therapeutic strategies and future prospects in cancer. Journal for Immunotherapy of Cancer. 2021;**9**

[152] Zhou T, Zhang LY, He JZ, Miao ZM, Li YY, Zhang YM, et al. Review: Mechanisms and perspective treatment of radioresistance in non-small cell lung cancer. Frontiers in Immunology. 2023;**14**:1133899

[153] Huang Z, Xiao Z, Yu L, Liu J, Yang Y, Ouyang W. Tumor-associated macrophages in non-small-cell lung cancer: From treatment resistance mechanisms to therapeutic targets. Critical Reviews in Oncology/Hematology. 2024;**196**

[154] Patysheva M, Frolova A, Larionova I, Afanasev S, Tarasova A, Cherdyntseva N, et al. Monocyte programming by cancer therapy. Frontiers in Immunology. 2022;**13**:994319

[155] Bergerud KMB, Berkseth M, Pardoll DM, Ganguly S, Kleinberg LR, Lawrence J, et al. Radiation therapy and myeloid-derived suppressor cells: Breaking down their cancerous partnership. International Journal of Radiation Oncology, Biology, Physics. 2024;**119**:42-55

[156] Genard G, Lucas S, Michiels C. Reprogramming of tumor-associated macrophages with anticancer therapies: Radiotherapy versus chemo- and immunotherapies. Frontiers in Immunology. 2017;**8**

[157] Brown JM, Thomas R, Nagpal S, Recht L. Macrophage exclusion after radiation therapy (MERT): A new and effective way to increase the therapeutic ratio of radiotherapy. Radiotherapy and Oncology. 2020;**144**:159-164

[158] Beach C, MacLean D, Majorova D, Arnold JN, Olcina MM. The effects of radiation therapy on the macrophage response in cancer. Frontiers in Oncology. 2022;**12**

[159] Becherini C, Lancia A, Detti B, Lucidi S, Scartoni D, Ingrosso G, et al. Modulation of tumor-associated macrophage activity with radiation therapy: A systematic review. Strahlentherapie und Onkologie. 2023;**199**:1173-1190

[160] Du J, Kageyama SI, Hirata H, Motegi A, Nakamura M, Hirano Y, et al. Comparative analysis of the immune responses in cancer cells irradiated with X-ray, proton and carbon-ion beams. Biochemical and Biophysical Research Communications. 2021;**585**:55-60

[161] Lu Z, Zheng X, Ding C, Zou Z, Liang Y, Zhou Y, et al. Deciphering the biological effects of radiotherapy in cancer cells. Biomolecules. 2022;**12**

[162] Vozenin MC, Bourhis J, Durante M. Towards clinical translation of FLASH radiotherapy. Nature Reviews. Clinical Oncology. 2022;**19**:791-803

[163] Sudo M, Tsutsui H, Fujimoto J. Carbon ion irradiation activates anti-cancer immunity. International Journal of Molecular Sciences. 2024;**25**

[164] Alaghband Y, Allen BD, Kramár EA, Zhang R, Drayson OGG, Ru N, et al. Uncovering the protective neurologic mechanisms of hypofractionated FLASH radiotherapy. Cancer Research Communications. 2023;**3**:725-737

[165] Almeida A, Godfroid C, Leavitt RJ, Montay-Gruel P, Petit B, Romero J, et al. Antitumor effect by either FLASH or conventional dose rate irradiation involves equivalent immune responses. International Journal of Radiation Oncology, Biology, Physics. 2024;**118**:1110-1122

[166] Montay-Gruel P, Bouchet A, Jaccard M, Patin D, Serduc R, Aim W, et al. X-rays can trigger the FLASH effect: Ultra-high dose-rate synchrotron light source prevents normal brain injury after whole brain irradiation in mice. Radiotherapy and Oncology. 2018;**129**:582-588

[167] Montay-Gruel P, Acharya MM, Petersson K, Alikhani L, Yakkala C, Allen BD, et al. Long-term neurocognitive benefits of FLASH radiotherapy driven by reduced reactive oxygen species. Proceedings of the National Academy of Sciences of the United States of America. 2019;**166**:10943-10951

[168] Montay-Gruel P, Markarian M, Allen BD, Baddour JD, Giedzinski E, Jorge PG, et al. Ultra-high-dose-rate FLASH irradiation limits reactive gliosis in the brain. Radiation Research. 2020;**194**:636-645

[169] Simmons DA, Lartey FM, Schüler E, Rafat M, King G, Kim A, et al. Reduced cognitive deficits after FLASH irradiation of whole mouse brain are associated with less hippocampal dendritic spine loss and neuroinflammation. Radiotherapy and Oncology. 2019;**139**:4-10

[170] Alaghband Y, Cheeks SN, Allen BD, Montay-Gruel P, Doan NL, Petit B, et al. Neuroprotection of radiosensitive juvenile mice by ultra-high dose rate FLASH irradiation. Cancers (Basel). 2020;**12**:1671

[171] Dokic I, Meister S, Bojcevski J, Tessonnier T, Walsh D, Knoll M, et al. Neuroprotective effects of ultra-high dose rate FLASH Bragg peak proton irradiation. International Journal of Radiation Oncology, Biology, Physics. 2022;**113**:614-623

[172] Zhu H, Xie D, Yang Y, Huang S, Gao X, Peng Y, et al. Radioprotective effect of X-ray abdominal FLASH irradiation: Adaptation to oxidative damage and inflammatory response may

be benefiting factors. Medical Physics. 2022;**49**:4812-4822

[173] Lim TL, Morral C, Verginadis II, Kim K, Luo L, Foley CJ, et al. Early inflammation and interferon signaling direct enhanced intestinal crypt regeneration after proton FLASH radiotherapy. BioRxiv. 2024

[174] Savage T, Pandey S, Guha C. Postablation modulation after single high-dose radiation therapy improves tumor control via enhanced immunomodulation. Clinical Cancer Research. 2020;**26**:910-921

[175] Shukla S, Saha T, Rama N, Acharya A, Le T, Bian F, et al. Ultra-high dose-rate proton FLASH improves tumor control. Radiotherapy and Oncology. 2023;**186**

[176] Fernandez-Palomo C, Trappetti V, Potez M, Pellicioli P, Krisch M, Laissue J, et al. Complete remission of mouse melanoma after temporally fractionated microbeam radiotherapy. Cancers (Basel). 2020;**12**:1-13

[177] Eggold JT, Chow S, Melemenidis S, Wang J, Natarajan S, Loo PE, et al. Abdominopelvic FLASH irradiation improves PD-1 immune checkpoint inhibition in preclinical models of ovarian cancer. Molecular Cancer Therapeutics. 2022;**21**:371-381

[178] Zhu H, Xie D, Wang Y, Huang R, Chen X, Yang Y, et al. Comparison of intratumor and local immune response between MV X-ray FLASH and conventional radiotherapies. Clinical and Translational Radiation Oncology. 2023;**38**:138-146

[179] Iturri L, Bertho A, Lamirault C, Juchaux M, Gilbert C, Espenon J, et al. Proton FLASH radiation therapy and immune infiltration: Evaluation in an orthotopic glioma rat model. International Journal of Radiation Oncology, Biology, Physics. 2023;**116**:655-665

[180] Padilla O, Minns HE, Wei HJ, Fan W, Webster-Carrion A, Tazhibi M, et al. Immune response following FLASH and conventional radiation in diffuse midline glioma. International Journal of Radiation Oncology, Biology, Physics. 2024;**119**:1248-1260

[181] Grabham P, Hu B, Sharma P, Geard C. Effects of ionizing radiation on three-dimensional human vessel models: Differential effects according to radiation quality and cellular development. Radiation Research. 2011;**175**:21-28

[182] Song CW, Payne JT, Levitt SH. Vascularity and blood flow in x-irradiated Walker carcinoma 256 of rats. Radiology. 1972;**104**:693-697

[183] Kuwahara Y, Mori M, Kitahara S, Fukumoto M, Ezaki T, Mori S, et al. Targeting of tumor endothelial cells combining 2 Gy/day of X-ray with everolimus is the effective modality for overcoming clinically relevant radioresistant tumors. Cancer Medicine. 2014;**3**:310-321

[184] Maeda A, Chen Y, Bu J, Mujcic H, Wouters BG, DaCosta RS. In vivo imaging reveals significant tumor vascular dysfunction and increased tumor hypoxia-inducible factor-1α expression induced by high single-dose irradiation in a pancreatic tumor model. International Journal of Radiation Oncology, Biology, Physics. 2017;**97**:184-194

[185] Zhou Q, Jin X, Zhao Y, Wang Y, Tao M, Cao Y, et al. Melanoma-associated fibroblasts in tumor-promotion flammation and antitumor immunity: Novel mechanisms

and potential immunotherapeutic strategies. Human Molecular Genetics. 2024;**33**:1186-1193

[186] Kato R, Haratani K, Hayashi H, Sakai K, Sakai H, Kawakami H, et al. Nintedanib promotes antitumour immunity and shows antitumour activity in combination with PD-1 blockade in mice: Potential role of cancer-associated fibroblasts. British Journal of Cancer. 2021;**124**:914-924

[187] Barcellos-Hoff MH. Stromal mediation of radiation carcinogenesis. Journal of Mammary Gland Biology and Neoplasia. 2010;**15**:381

[188] Hellevik T, Pettersen I, Berg V, Bruun J, Bartnes K, Busund LT, et al. Changes in the secretory profile of NSCLC-associated fibroblasts after ablative radiotherapy: Potential impact on angiogenesis and tumor growth. Translational Oncology. 2013;**6**:66

[189] Hellevik T, Pettersen I, Berg V, Winberg JO, Moe BT, Bartnes K, et al. Cancer-associated fibroblasts from human NSCLC survive ablative doses of radiation but their invasive capacity is reduced. Radiation Oncology. 2012;7

[190] Bliley R, Avant A, Medina TM, Lanning RM. Radiation and melanoma: Where are we now? Current Oncology Reports. 2024;**26**:904-914

[191] Rogers SJ, Puric E, Eberle B, Datta NR, Bodis SB. Radiotherapy for melanoma: More than DNA damage. Dermatology Research and Practice. 2019;**2019**:9435389

[192] Strojan P. Role of radiotherapy in melanoma management. Radiology and Oncology. 2010;**44**:1

[193] Shi W. Radiation therapy for melanoma. In: Cutaneous Melanoma: Etiology and Therapy. Codon Publications; 2017. pp. 101-120

[194] Domingues B, Lopes JM, Soares P, Pópulo H. Melanoma treatment in review. ImmunoTargets and Therapy. 2018;**7**:35-49

[195] Dhanyamraju PK, Patel TN. Melanoma therapeutics: A literature review. Journal of Biomedical Research. 2022;**36**:77

[196] Hawkins BS. The COMS randomized trial of iodine 125 brachytherapy for choroidal melanoma: V. Twelve-year mortality rates and prognostic factors: COMS report No. 28. Archives of Ophthalmology. 2006;**124**:1684-1693

[197] Seibel I, Cordini D, Rehak M, Hager A, Riechardt AI, Böker A, et al. Local recurrence after primary proton beam therapy in uveal melanoma: Risk factors, retreatment approaches, and outcome. American Journal of Ophthalmology. 2015;**160**:628-636

[198] Farshad A, Burg G, Panizzon R, Dummer R. A retrospective study of 150 patients with lentigo maligna and lentigo maligna melanoma and the efficacy of radiotherapy using grenz or soft X-rays. The British Journal of Dermatology. 2002;**146**:1042-1046

[199] Krengli M, Jereczek-Fossa BA, Kaanders JHAM, Masini L, Beldì D, Orecchia R. What is the role of radiotherapy in the treatment of mucosal melanoma of the head and neck? Critical Reviews in Oncology/Hematology. 2008;**65**:121-128

[200] Varey AHR, Goumas C, Hong AM, Mann GJ, Fogarty GB, Stretch JR, et al. Neurotropic melanoma: An analysis of the clinicopathological features, management strategies and survival

outcomes for 671 patients treated at a tertiary referral center. Modern Pathology. 2017;**30**:1538-1550

[201] Henderson MA, Burmeister BH, Ainslie J, Fisher R, Di Iulio J, Smithers BM, et al. Adjuvant lymph-node field radiotherapy versus observation only in patients with melanoma at high risk of further lymph-node field relapse after lymphadenectomy (ANZMTG 01.02/TROG 02.01): 6-year follow-up of a phase 3, randomised controlled trial. The Lancet Oncology. 2015;**16**:1049-1060

[202] Horvat AG, Kovač V, Strojan P. Radiotherapy in palliative treatment of painful bone metastases. Radiology and Oncology. 2009;**43**

[203] Fife KM, Colman MH, Stevens GN, Firth IC, Moon D, Shannon KF, et al. Determinants of outcome in melanoma patients with cerebral metastases. Journal of Clinical Oncology. 2004;**22**:1293-1300

[204] Vanpouille-Box C, Formenti SC, Demaria S. Toward precision radiotherapy for use with immune checkpoint blockers. Clinical Cancer Research. 2018;**24**:259-265

[205] Escorcia FE, Postow MA, Barker CA. Radiotherapy and immune checkpoint blockade for melanoma: A promising combinatorial strategy in need of further investigation. Cancer Journal. 2017;**23**:32

[206] Tagliaferri L, Lancellotta V, Fionda B, Mangoni M, Casà C, Di Stefani A, et al. Immunotherapy and radiotherapy in melanoma: A multidisciplinary comprehensive review. Human Vaccines & Immunotherapeutics. 2021;**18**:1903827

[207] Gupta A, Roberts C, Tysoe F, Goff M, Nobes J, Lester J, et al. RADVAN: A randomised phase 2 trial of WBRT plus vandetanib for melanoma brain metastases – Results and lessons learnt. British Journal of Cancer. 2016;**115**:1193

[208] Schmidt-Ullrich RK, Johnson CR. Role of radiotherapy and hyperthermia in the management of malignant melanoma. Seminars in Surgical Oncology. 1996;**12**:407-415

[209] Overgaard J, Bentzen SM, Gonzalez Gonzalez D, Hulshof MCCM, Arcangeli G, Dahl O, et al. Randomised trial of hyperthermia as adjuvant to radiotherapy for recurrent or metastatic malignant melanoma. European Society for Hyperthermic Oncology. Lancet. 1995;**345**:540-543

[210] Kim J, Hahn E, Ahmed S. Combination hyperthermia and radiation therapy for malignant melanoma. Cancer. 1982;**50**:478-482

[211] Kwon S, Jung S, Baek SH. Combination therapy of radiation and hyperthermia, focusing on the synergistic anti-cancer effects and research trends. Antioxidants. 2023;**12**:924

[212] Samant RS, Gerig L, Montgomery L, MacRae R, Fox G, Nyiri B, et al. The emerging role of IG-IMRT for palliative radiotherapy: A single-institution experience. Current Oncology. 2009;**16**:40

[213] Anderson ES, Postow MA, Wolchok JD, Young RJ, Ballangrud Å, Chan TA, et al. Melanoma brain metastases treated with stereotactic radiosurgery and concurrent pembrolizumab display marked regression; efficacy and safety of combined treatment. Journal for Immunotherapy of Cancer. 2017;**5**

[214] Liermann J, Winkler JK, Syed M, Neuberger U, Reuss D, Harrabi S,

et al. Stereotactic radiosurgery with concurrent immunotherapy in melanoma brain metastases is feasible and effective. Frontiers in Oncology. 2020;**10**

[215] Chu SA, Chen TB, Chen HJ, Wang KW, Chen JS, Chuang FC, et al. Gamma knife radiosurgery as a salvage treatment for nasopharyngeal carcinoma with skull base and intracranial invasion (T4b). Life (Basel). 2022;**12**

[216] Trentesaux V, Maiezza S, Bogart E, Le Deley MC, Meyer E, Vanquin L, et al. Stereotactic body radiotherapy as a viable treatment on extracranial oligometastases in melanoma patients: A retrospective multicentric study. Frontiers in Oncology. 2024;**14**

[217] Curti B, Crittenden M, Seung SK, Fountain CB, Payne R, Chang SC, et al. Randomized phase II study of stereotactic body radiotherapy and interleukin-2 versus interleukin-2 in patients with metastatic melanoma. Journal for Immunotherapy of Cancer. 2020;**8**

[218] Jenkins SV, Johnsrud AJ, Dings RPM, Griffin RJ. Bystander effects in spatially fractionated radiation therapy: From molecule to organism to clinical implications. Seminars in Radiation Oncology. 2024;**34**:284-291

[219] Dilmanian FA, Qu Y, Feinendegen LE, Peña LA, Bacarian T, Henn FA, et al. Tissue-sparing effect of x-ray microplanar beams particularly in the CNS: Is a bystander effect involved? Experimental Hematology. 2007;**35**:69-77

[220] Marques L, Costa B, Pereira M, Silva A, Santos J, Saldanha L, et al. Advancing precision medicine: A review of innovative In Silico approaches for drug development, clinical pharmacology and personalized healthcare. Pharmaceutics. 2024;**16**

[221] Li Q, Lei X, Zhu J, Zhong Y, Yang J, Wang J, et al. Radiotherapy/chemotherapy-immunotherapy for cancer management: From mechanisms to clinical implications. Oxidative Medicine and Cellular Longevity. 2023;**2023**:7530794

[222] Ridolfi L, de Rosa F, Ridolfi R, Gentili G, Valmorri L, Scarpi E, et al. Radiotherapy as an immunological booster in patients with metastatic melanoma or renal cell carcinoma treated with high-dose Interleukin-2: Evaluation of biomarkers of immunologic and therapeutic response. Journal of Translational Medicine. 2014;**12**

[223] de Rosa F, Ridolfi L, Ridolfi R, Gentili G, Valmorri L, Nanni O, et al. Vaccination with autologous dendritic cells loaded with autologous tumor lysate or homogenate combined with immunomodulating radiotherapy and/or preleukapheresis IFN-α in patients with metastatic melanoma: A randomised "proof-of-principle" phase II study. Journal of Translational Medicine. 2014;**12**

[224] Hamanishi J, Mandai M, Ikeda T, Minami M, Kawaguchi A, Murayama T, et al. Safety and antitumor activity of anti-PD-1 antibody, nivolumab, in patients with platinum-resistant ovarian cancer. Journal of Clinical Oncology. 2015;**33**:4015-4022

[225] Varga A, Piha-Paul SA, Ott PA, Mehnert JM, Berton-Rigaud D, Johnson EA, et al. Antitumor activity and safety of pembrolizumab in patients (pts) with PD-L1 positive advanced ovarian cancer: Interim results from a phase Ib study. Journal of Clinical Oncology. 2015;**33**:5510-5510

[226] Weber JS, Yang JC, Atkins MB, Disis ML. Toxicities of immunotherapy

for the practitioner. Journal of Clinical Oncology. 2015;**33**:2092-2099

[227] Moore KN, Bookman M, Sehouli J, Miller A, Anderson C, Scambia G, et al. Atezolizumab, bevacizumab, and chemotherapy for newly diagnosed stage III or IV ovarian cancer: Placebo-controlled randomized phase III trial (IMagyn050/GOG 3015/ENGOT-OV39). Journal of Clinical Oncology. 2021;**39**:1842-1855

[228] Pujade-Lauraine E, Fujiwara K, Ledermann JA, Oza AM, Kristeleit R, Ray-Coquard IL, et al. Avelumab alone or in combination with chemotherapy versus chemotherapy alone in platinum-resistant or platinum-refractory ovarian cancer (JAVELIN ovarian 200): An open-label, three-arm, randomised, phase 3 study. The Lancet Oncology. 2021;**22**:1034-1046

[229] Khatoon E, Parama D, Kumar A, Alqahtani MS, Abbas M, Girisa S, et al. Targeting PD-1/PD-L1 axis as new horizon for ovarian cancer therapy. Life Sciences. 2022;**306**

[230] Turner TB, Buchsbaum DJ, Straughn JM, Randall TD, Arend RC. Ovarian cancer and the immune system - the role of targeted therapies. Gynecologic Oncology. 2016;**142**:349-356

[231] Baert T, Vergote I, Coosemans A. Ovarian cancer and the immune system. Gynecologic Oncology Reports. 2017;**19**:57

[232] Deng L, Liang H, Xu M, Yang X, Burnette B, Arina A, et al. STING-dependent cytosolic DNA sensing promotes radiation-induced type I interferon-dependent antitumor immunity in immunogenic tumors. Immunity. 2014;**41**:843-852

[233] Matsumura S, Wang B, Kawashima N, Braunstein S, Badura M, Cameron TO, et al. Radiation-induced CXCL16 release by breast cancer cells attracts effector T cells. Journal of Immunology. 2008;**181**:3099

[234] Vanpouille-Box C, Alard A, Aryankalayil MJ, Sarfraz Y, Diamond JM, Schneider RJ, et al. DNA exonuclease Trex1 regulates radiotherapy-induced tumour immunogenicity. Nature Communications. 2017;**8**:15618

[235] Burnette BC, Liang H, Lee Y, Chlewicki L, Khodarev NN, Weichselbaum RR, et al. The efficacy of radiotherapy relies upon induction of type i interferon-dependent innate and adaptive immunity. Cancer Research. 2011;**71**:2488-2496

[236] Herrera FG, Bourhis J, Coukos G. Radiotherapy combination opportunities leveraging immunity for the next oncology practice. CA: a Cancer Journal for Clinicians. 2017;**67**:65-85

[237] Fields EC, McGuire WP, Lin L, Temkin SM. Radiation treatment in women with ovarian cancer: Past, present, and future. Frontiers in Oncologia. 2017:7

[238] Herrera FG, Irving M, Kandalaft LE, Coukos G. Rational combinations of immunotherapy with radiotherapy in ovarian cancer. The Lancet Oncology. 2019;**20**:e417-e433

[239] Fyles AW, Dembo AJ, Bush RS, Levin W, Manchul LA, Pringle JF, et al. Analysis of complications in patients treated with abdomino-pelvic radiation therapy for ovarian carcinoma. International Journal of Radiation Oncology, Biology, Physics. 1992;**22**:847-851

[240] Sologuren I, Rodríguez-Gallego C, Lara PC. Immune effects of high dose radiation treatment: Implications of

ionizing radiation on the development of bystander and abscopal effects. Translational Cancer Research. 2014;**3**:18-31

[241] Herrera FG, Ronet C, de Olza MO, Barras D, Crespo I, Andreatta M, et al. Low-dose radiotherapy reverses tumor immune desertification and resistance to immunotherapy. Cancer Discovery. 2022;**12**:108-133

[242] Mole RH. Whole body irradiation; radiobiology or medicine? The British Journal of Radiology. 1953;**26**:234-241

[243] Sima LE, Matei D, Condello S. The outside-in journey of tissue transglutaminase in cancer. Cells. 2022;**11**:1779

[244] Gundemir S, Monteagudo A, Akbar A, Keillor JW, Johnson GVW. The complex role of transglutaminase 2 in glioblastoma proliferation. Neuro-Oncology. 2017;**19**:208-218

[245] Satpathy M, Cao L, Pincheira R, Emerson R, Bigsby R, Nakshatri H, et al. Enhanced peritoneal ovarian tumor dissemination by tissue transglutaminase. Cancer Research. 2007;**67**:7194-7202

[246] Iacobuzio-Donahue CA, Ashfaq R, Maitra A, Adsay N, GKB S-O, Hollingsworth M, et al. Highly expressed genes in pancreatic ductal adenocarcinomas: A comprehensive characterization and comparison of the transcription profiles obtained from three major technologies. Cancer Research. 2003;**63**:8614-8622

[247] Choi CM, Jang SJ, Park SY, Choi YB, Jeong JH, Kim DS, et al. Transglutaminase 2 as an independent prognostic marker for survival of patients with non-adenocarcinoma subtype of non-small cell lung cancer. Molecular Cancer. 2011;**10**:119

[248] Grigoriev MY, Suspitsin EN, Togo AV, Pozharisski KM, Ivanova OA, Nardacci R, et al. Tissue transglutaminase expression in breast carcinomas. Journal of Experimental & Clinical Cancer Research. 2001;**20**:265-268

[249] Jee YH, Mangala LS, Fok JY, Lin YG, Merritt WM, Spannuth WA, et al. Clinical and biological significance of tissue transglutaminase in ovarian carcinoma. Cancer Research. 2008;**68**:5849-5858

[250] Jeong JH, Cho BC, Shim HS, Kim HR, Lim SM, Kim SK, et al. Transglutaminase 2 expression predicts progression free survival in non-small cell lung cancer patients treated with epidermal growth factor receptor tyrosine kinase inhibitor. Journal of Korean Medical Science. 2013;**28**:1005-1014

[251] Verma A, Wang H, Manavathi B, Fok JY, Mann AP, Kumar R, et al. Increased expression of tissue transglutaminase in pancreatic ductal adenocarcinoma and its implications in drug resistance and metastasis. Cancer Research. 2006;**66**:10525-10533

[252] Lei X, Cao K, Chen Y, Shen H, Liu Z, Qin H, et al. Nuclear transglutaminase 2 interacts with topoisomerase IIα to promote DNA damage repair in lung cancer cells. Journal of Experimental & Clinical Cancer Research. 2021;**40**

[253] Aepler J, Wodtke J, Wodtke R, Haase-Kohn C, Löser R, Pietzsch J, et al. The role of transglutaminase 2 in the radioresistance of melanoma cells. Cells. 2022;**11**:1342

[254] Zheng W, Chen Q, Liu H, Zeng L, Zhou Y, Liu X, et al. SDC1-dependent TGM2 determines radiosensitivity in glioblastoma by

coordinating EPG5-mediated fusion of autophagosomes with lysosomes. Autophagy. 2022;**19**:839

[255] Berg TJ, Marques C, Pantazopoulou V, Johansson E, Von Stedingk K, Lindgren D, et al. The irradiated brain microenvironment supports glioma stemness and survival via astrocyte-derived transglutaminase 2. Cancer Research. 2021;**81**:2101-2115

[256] Yin J, Oh YT, Kim JY, Kim SS, Choi E, Kim TH, et al. Transglutaminase 2 inhibition reverses mesenchymal transdifferentiation of glioma stem cells by regulating C/EBPβ signaling. Cancer Research. 2017;**77**:4973-4984

[257] Kausar T, Sharma R, Hasan MR, Tripathi SC, Saraya A, Chattopadhyay TK, et al. Clinical significance of GPR56, transglutaminase 2, and NF-κB in esophageal squamous cell carcinoma. Cancer Investigation. 2011;**29**:42-48

[258] Leicht DT, Kausar T, Wang Z, Ferrer-Torres D, Wang TD, Thomas DG, et al. TGM2: A cell surface marker in esophageal adenocarcinomas. Journal of Thoracic Oncology. 2014;**9**:872-881

[259] Jin T, Lin HX, Lin H, Guo LB, Ge N, Cai XY, et al. Expression TGM2 and BNIP3 have prognostic significance in laryngeal cancer patients receiving surgery and postoperative radiotherapy: A retrospective study. Journal of Translational Medicine. 2012;**10**:1-9

[260] Belkin AM. Extracellular TG2: Emerging functions and regulation. The FEBS Journal. 2011;**278**:4704-4716

[261] Sestito C, Brevé JJP, van Eggermond MCJA, Killestein J, Teunissen CE, van Rossum J, et al. Monocyte-derived tissue transglutaminase in multiple sclerosis patients: Reflecting an anti-inflammatory status and function of the cells? Journal of Neuroinflammation. 2017;**14**:1-12

[262] Khanna M, Chelladurai B, Gavini A, Li L, Shao M, Courtney D, et al. Targeting ovarian tumor cell adhesion mediated by tissue transglutaminase. Molecular Cancer Therapeutics. 2011;**10**:626-636

[263] Yakubov B, Chen L, Belkin AM, Zhang S, Chelladurai B, Zhang ZY, et al. Small molecule inhibitors target the tissue transglutaminase and fibronectin interaction. PLoS One. 2014;**9**:e89285

[264] Sima LE, Yakubov B, Zhang S, Condello S, Grigorescu AA, Nwani NG, et al. Small molecules target the interaction between tissue transglutaminase and fibronectin. Molecular Cancer Therapeutics. 2019;**18**:1057-1068

[265] Badarau E, Wang Z, Rathbone DL, Costanzi A, Thibault T, Murdoch CE, et al. Development of potent and selective tissue transglutaminase inhibitors: Their effect on TG2 function and application in pathological conditions. Chemistry & Biology. 2015;**22**:1347-1361

[266] Akbar A, McNeil NMR, Albert MR, Ta V, Adhikary G, Bourgeois K, et al. Structure-activity relationships of potent, targeted covalent inhibitors that abolish both the transamidation and GTP binding activities of human tissue transglutaminase. Journal of Medicinal Chemistry. 2017;**60**:7910-7927

[267] Gallo M, Ferrari E, Terrazzan A, Brugnoli F, Spisni A, Taccioli C, et al. Metabolic characterisation of transglutaminase 2 inhibitor effects in breast cancer cell lines. The FEBS Journal. 2023;**290**:5411-5433

[268] Huayin S, Dong Y, Chihong Z, Xiaoqian Q, Danying W, Jianguo F. Transglutaminase 2 inhibitor KCC009 induces p53-independent radiosensitization in lung adenocarcinoma cells. Medical Science Monitor. 2016;**22**:5041-5048

[269] Sima LE, Chen S, Cardenas H, Zhao G, Wang Y, Ivan C, et al. Loss of host tissue transglutaminase boosts antitumor T cell immunity by altering STAT1/STAT3 phosphorylation in ovarian cancer. Journal for Immunotherapy of Cancer. 2021;**9**:e002682

[270] Wang Z, Moresco P, Yan R, Li J, Gao Y, Biasci D, et al. Carcinomas assemble a filamentous CXCL12-keratin-19 coating that suppresses T cell-mediated immune attack. Proceedings of the National Academy of Sciences of the United States of America. 2022;**119**:e2119463119